Macmillan/McGraw-Hill • Glencoe

California Math Triumphs

Volume 4A

The Core Processes of Mathematics

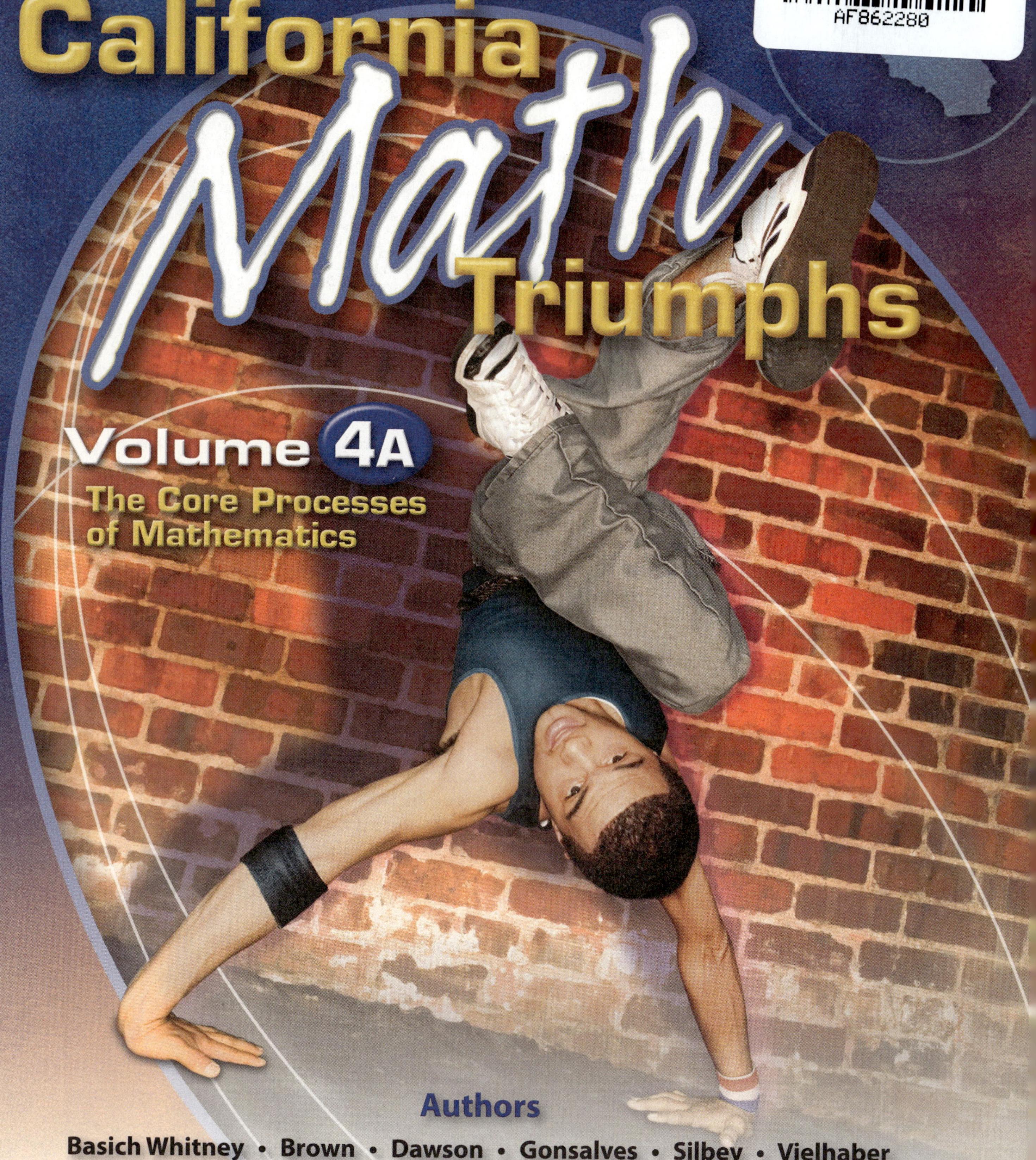

Authors

Basich Whitney • Brown • Dawson • Gonsalves • Silbey • Vielhaber

Macmillan/McGraw-Hill
Glencoe

Photo Credits

Cover Thinkstock/Alamy; **iv** (tl)(tc)(tr)The McGraw-Hill Companies, (cl)(c)Doug Martin, (cr)Aaron Haupt, (bl)(bc)The McGraw-Hill Companies; **v** The McGraw-Hill Companies; **vii** Ian Grant/Alamy; **viii** Medioimages/PunchStock; **ix** Digital Vision/PunchStock; **x 1** CORBIS; **2–3** Brian Pieters/Masterfile; **10** Mike Powell/Getty Images; **14** Michael Houghton/StudiOhio; **16** The McGraw-Hill Companies; **40–41** F. Lukasseck/Masterfile.

The McGraw·Hill Companies

Macmillan/McGraw-Hill
Glencoe

Send all inquiries to:
Glencoe/McGraw-Hill
8787 Orion Place
Columbus, OH 43240-4027

ISBN: 978-0-07-878209-1
MHID: 0-07-878209-0

California Math Triumphs
Volume 4A

Printed in the United States of America.

5 6 7 8 9 10 066 16 15 14 13 12 11 10 09

California Math Triumphs

Authors and Consultants

AUTHORS

Frances Basich Whitney
Project Director, Mathematics K–12
Santa Cruz County Office of Education
Capitola, California

Kathleen M. Brown
Math Curriculum Staff Developer
Washington Middle School
Long Beach, California

Dixie Dawson
Math Curriculum Leader
Long Beach Unified
Long Beach, California

Philip Gonsalves
Mathematics Coordinator
Alameda County Office of Education
Hayward, California

Robyn Silbey
Math Specialist
Montgomery County Public Schools
Gaithersburg, Maryland

Kathy Vielhaber
Mathematics Consultant
St. Louis, Missouri

CONTRIBUTING AUTHORS

Viken Hovsepian
Professor of Mathematics
Rio Hondo College
Whittier, California

Dinah Zike
Educational Consultant,
Dinah-Might Activities, Inc.
San Antonio, Texas

CONSULTANTS

Assessment

Donna M. Kopenski, Ed.D.
Math Coordinator K-5
City Heights Educational Collaborative
San Diego, California

Instructional Planning and Support

Beatrice Luchin
Mathematics Consultant
League City, Texas

ELL Support and Vocabulary

ReLeah Cossett Lent
Author/Educational Consultant
Alford, Florida

California Advisory Board

CALIFORNIA ADVISORY BOARD

Glencoe wishes to thank the following professionals for their invaluable feedback during the development of the program. They reviewed the table of contents, the prototype of the Student Study Guide, the prototype of the Teacher Wraparound Edition, and the professional development plan.

Linda Anderson
4th/5th Grade Teacher
Oliveira Elementary School,
Fremont, California

Cheryl L. Avalos
Mathematics Consultant
Retired Teacher
Hacienda Heights, California

Bonnie Awes
Teacher, 6th Grade Math
Monroe Clark Middle School
San Diego, California

Kathleen M. Brown
Math Curriculum Staff Developer
Washington Middle School
Long Beach, California

Carol Cronk
Mathematics Program Specialist
San Bernardino City Unified School District
San Bernardino, California

Audrey M. Day
Classroom Teacher
Rosa Parks Elementary School
San Diego, California

Jill Fetters
Math Teacher
Tevis Jr. High School
Bakersfield, California

Grant A. Fraser, Ph.D.
Professor of Mathematics
California State University, Los Angeles
Los Angeles, California

Eric Kimmel
Mathematics Department Chair
Frontier High School
Bakersfield, California

Donna M. Kopenski, Ed.D.
Math Coordinator K–5
City Heights Educational Collaborative
San Diego, California

Michael A. Pease
Instructional Math Coach
Aspire Public Schools
Oakland, California

Chuck Podhorsky, Ph.D.
Math Director
City Heights Educational Collaborative
San Diego, California

Arthur K. Wayman, Ph.D.
Professor Emeritus
California State University, Long Beach
Long Beach, California

Frances Basich Whitney
Project Director, Mathematics K–12
Santa Cruz County Office of Education
Capitola, CA

Mario Borrayo
Teacher
Rosa Parks Elementary
San Diego, California

Melissa Bray
K–8 Math Resource Teacher
Modesto City Schools
Modesto, California

California Reviewers

CALIFORNIA REVIEWERS

Each California Reviewer reviewed at least two chapters of the Student Study Guides, providing feedback and suggestions for improving the effectiveness of the mathematics instruction.

Bobbi Anne Barnowsky
Math Teacher
California College Preparatory Academy
Oakland, California

Eppie Leamy Chung
Teacher
Modesto City Schools
Modesto, California

Judy Descoteaux
Mathematics Teacher
Thornton Junior High School
Fremont, California

Paul J. Fogarty
Mathematics Lead
Aspire Public Schools
Modesto, California

Lisa Majarian
Classroom Teacher
Cottonwood Creek Elementary
Visalia, California

Melody McGuire
6th and 7th Grade Math Teacher
McKinleyville Middle School
McKinleyville, California

Monica S. Patterson
Educator
Aspire Public Schools
Modesto, California

Rechelle Pearlman
4th Grade Teacher
Wanda Hirsch Elementary School
Tracy, California

Armida Picon
5th Grade Teacher
Mineral King School
Visalia, California

Anthony J. Solina
Lead Educator
Aspire Public Schools
Stockton, California

Volume 4A The Core Processes of Mathematics

Chapter 1 Operations and Equality

Chapters 1 and 2 are contained in Volume 4A. Chapters 3, 4, and 5 are contained in Volume 4B.

Standards Addressed in This Chapter

3AF1.0 Students select appropriate symbols, operations, and properties to represent, describe, simplify, and solve simple number relationships.

4AF1.1 Use letters, boxes, or other symbols to stand for any number in simple expressions or equations (e.g., demonstrate an understanding and the use of the concept of a variable).

4AF2.1 Know and understand that equals added to equals are equal.

4AF2.2 Know and understand that equals multiplied by equals are equal.

Point Lobos State Park

Contents

Chapter 2 Math Fundamentals

Standards Addressed in This Chapter

2AF1.1 Use the commutative and associative rules to simplify mental calculations and to check results.

3AF1.5 Recognize and use the commutative and associative properties of multiplication (e.g., if $5 \times 7 = 35$, then what is 7×5? and if $5 \times 7 \times 3 = 105$, then what is $7 \times 3 \times 5$?).

5AF1.3 Know and use the distributive property in equations and expressions with variables.

7AF1.2 Use the correct order of operations to evaluate algebraic expressions such as $3(2x + 5)^2$.

Mustard plants in Napa Valley

Chapter 3 Math Expressions

Chapters 1 and 2 are contained in Volume 4A.
Chapters 3, 4, and 5 are contained in Volume 4B.

Standards Addressed in This Chapter

5AF1.2 Use a letter to represent an unknown number; write and evaluate simple algebraic expressions in one variable by substitution.

6AF1.2 Write and evaluate an algebraic expression for a given situation, using up to three variables.

7AF1.1 Use variables and appropriate operations to write an expression, an equation, an inequality, or a system of equations or inequalities that represents a verbal description (e.g., three less than a number, half as large as area A).

7AF1.3 Simplify numerical expressions by applying properties of rational numbers (e.g., identity, inverse, distributive, associative, commutative) and justify the process used.

Burney Falls

Contents

Chapter 4 Linear Equations

Standards Addressed in This Chapter

4AF2.1 Know and understand that equals added to equals are equal.

4AF2.2 Know and understand that equals multiplied by equals are equal.

7NS1.3 Convert fractions to decimals and percents and use these representations in estimations, computations, and applications.

7AF1.1 Use variables and appropriate operations to write an expression, an equation, an inequality, or a system of equations or inequalities that represents a verbal description (e.g., three less than a number, half as large as area A).

7AF4.0 Students solve simple linear equations and inequalities over the rational numbers.

Alabama Hills, Owens Valley

Contents

Chapter 5 Inequalities

Assessment

Standards Addressed in This Chapter

7NS1.3 Convert fractions to decimals and percents and use these representations in estimations, computations, and applications.

7AF1.1 Use variables and appropriate operations to write an expression, an equation, an inequality, or a system of equations or inequalities that represents a verbal description (e.g., three less than a number, half as large as area A).

7AF4.0 Students solve simple linear equations and inequalities over the rational numbers.

California poppies and gazanias

Chapter 1

Operations and Equality

When you shop, you figure out what you can buy.

For example, which shirt costs more? How much more does it cost? If you buy two shirts, will you have enough money left to buy a snack on the way home?

STEP 1 Quiz

Are you ready for Chapter 1? Take the Online Readiness Quiz at ca.mathtriumphs.com to find out.

STEP 2 Preview

Get ready for Chapter 1. Review these skills and compare them with what you'll learn in this chapter.

What You Know	What You Will Learn
You know how to add and subtract.	*Lesson 1-1*

What You Know

You know how to add and subtract.

Examples: $15 + 8 \Rightarrow$

$$\begin{array}{r} \scriptstyle 1 \\ 15 \\ +\ 8 \\ \hline 23 \end{array}$$

$100 - 16 \Rightarrow$

$$\begin{array}{r} \scriptstyle 9 \\ \scriptstyle 0\,\cancel{10}\,10 \\ \cancel{100} \\ -\ 16 \\ \hline 84 \end{array}$$

TRY IT!

1. $14 + 7 =$ ____
2. $9 + 17 =$ ____
3. $40 - 12 =$ ____
4. $54 - 5 =$ ____

What You Will Learn

Lesson 1-1

If you want to know the **total** amount, you add.

If you want to know **how much more** one item costs than another, you should subtract the prices.

What You Know

You know how to multiply and divide.

Examples: $3 \times 8 = 24$

$36 \div 9 = 4$

TRY IT!

5. $14 \times 2 =$ ____
6. $4 \times 17 =$ ____
7. $40 \div 5 =$ ____
8. $60 \div 6 =$ ____

What You Will Learn

Lesson 1-2

If you want to know the **total amount of several equal sets of items,** you multiply.

If you want to **separate a group of items into equal sets,** you divide.

Lesson 1-1

Addition and Subtraction Operations

3AF1.0 Students select appropriate symbols, operations, and properties to represent, describe, simplify, and solve simple number relationships.

KEY Concept

Operation	Answer	Key Words	Symbol
addition	sum	how many, total, in all, combined, together	+
subtraction	difference	how many more, how many left, less, more than	–

addition: $5 + 3 = 8$ — addends: 5 and 3; sum: 8

subtraction: $6 - 2 = 4$ — difference: 4

VOCABULARY

addition
an operation on two or more numbers that gives a total (*sum*)

addends
numbers being added together

sum
the answer to an *addition* problem

subtraction
an operation on two numbers that tells how many are left (*difference*) when some are taken away

difference
the answer to a *subtraction* problem

You can use *addition* to check your work on a *subtraction* problem. Use *subtraction* to check your work on an *addition* problem.

Example 1

Solve.

SCHOOL SUPPLIES Adam had 9 pencils. He gave 3 pencils to Carmen. How many pencils did Adam have left?

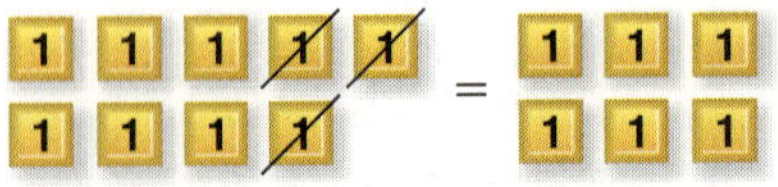

1. You are asked how many pencils were left after Adam gave some to Carmen.
2. Write a subtraction sentence.

pencils Adam began with	–	pencils he gave to Carmen	=	pencils Adam has left
9	–	3	=	6

3. Answer the question.
   Adam has 6 pencils left.

YOUR TURN!

Solve.

SPORTS Jase scored 3 points for his kickball team. Lia scored 5 points for her kickball team. How many more points did Lia score than Jase?

1. You are asked ________________ points Lia scored than Jase.
2. Write a ________________ sentence.

points Lia scored	−	points Jase scored	=	difference in points scored
______	−	______	=	______

3. Answer the question. ________________________________

Example 2

Solve.

Bonnie rode her bike for 20 minutes on Monday and 15 minutes on Tuesday. How many minutes did Bonnie ride in all?

1. You are asked how many minutes Bonnie rode her bike in all.
2. Write an addition sentence.

minutes on Monday	+	minutes on Tuesday	=	total number of minutes
20	+	15	=	35

3. Answer the question.
   Bonnie rode for 35 minutes in all.

YOUR TURN!

Solve.

Dina had 3 goldfish in her fish tank. Yesterday she bought 8 more. How many goldfish does Dina have altogether?

1. You are asked how many fish Dina has ________________.
2. Write an ________________ sentence.

fish Dina began with	+	fish she bought	=	total number of fish
______	+	______	=	______

3. Answer the question.
   ____

Who is Correct?

What operation can you use to find a total?

Ned	Pati	Jamaal
subtraction	**addition or subtraction**	**addition**

Circle correct answer(s). Cross out incorrect answer(s).

GO ON

Guided Practice

Name each operation modeled.

1 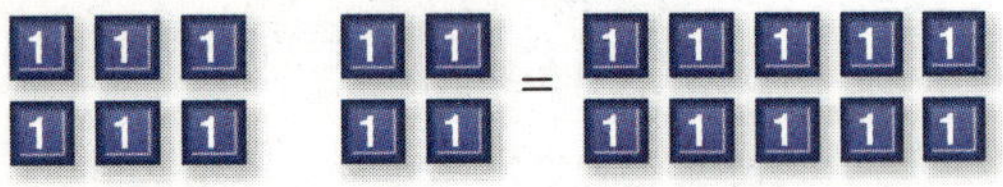

____

2 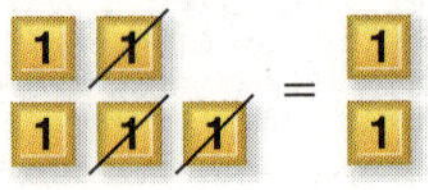

____

Step by Step Practice

Solve.

3 **CRAFTS** Tiara made spirit bows for the school carnival. She started with 85 feet of ribbon. She used 62 feet to create bows. How much ribbon did Tiara have left?

Step 1 You are asked how much ribbon was __________.

Step 2 Write a __________ sentence.
85 ______ 62 = ______

Step 3 Answer the question.
Tiara had ______ feet of ribbon left.

Name the operation needed to solve each problem. Write a number sentence to solve each problem. Answer the question.

4 **TRAVEL** Jamal drove from Los Angeles to Sacramento. He drove 150 miles before lunch and 250 miles after lunch. How many miles did Jamal drive in all?

____

5 **SCHOOL** Alberto took two geography tests. He earned a 73 on the first test and an 89 on the second test. How many more points did Alberto earn on the second test than on the first?

____

Step by Step Problem-Solving Practice

Problem-Solving Strategies
- ☑ Solve a simpler problem.
- ☐ Look for a pattern.
- ☐ Guess and check.
- ☐ Draw a diagram.
- ☐ Work backward.

Solve.

6 **CARNIVAL** You have 4 tickets to use at a carnival. You want to ride the Ferris wheel, but it takes 12 tickets. A friend gives you 5 tickets. How many more tickets do you need to ride the Ferris wheel?

Understand Read the problem. Write what you know.

You have ______ tickets.

A friend gives you ______ more tickets.

You need ______ tickets for the ride.

Plan Pick a strategy. One strategy is to solve a simpler problem. Two simpler problems are:

1. You have ______ tickets. Your friend gives you ______ more. How many tickets do you have in all?
2. The ride costs ______ tickets. How many more tickets do you need?

Solve Write a number sentence to solve the first problem.

1. ______ + ______ = ______

Write a number sentence to solve the second problem.

2. ______ − ______ = ______

You need ______ more tickets to ride the Ferris wheel.

Check Does your answer make sense? Draw a diagram to check your answer.

Draw the ______ tickets needed. Mark out the ______ tickets you have and the ______ tickets from your friend. Count the tickets not marked out.

You need ______ more tickets.

GO ON

Solve.

7 **PROJECTS** Mila has a 20-foot piece of rope. She needs two 9-foot pieces of rope to use for her science fair project. How much rope will Mila have left? ________ Check off each step.

_____ **Understand**

_____ **Plan**

_____ **Solve**

_____ **Check**

8 **MONEY** Andrea has $100 to spend on clothes. She spends $30 on a pair of jeans and $25 on a sweater. How much money does Andrea have left to spend?

____

9 **Reflect** Name one key word for addition and one for subtraction that will help you choose the operation to use.

____

Skills, Concepts, and Problem Solving

Name each operation modeled.

10 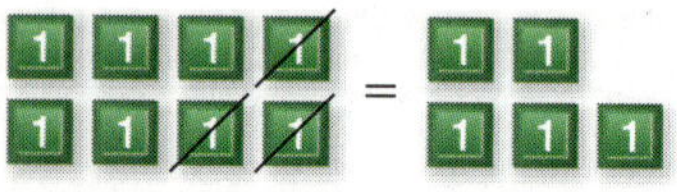

____

11 

____

Name the operation needed to solve each problem. Write a number sentence to solve each problem. Answer the question.

12 **WORK** Kathy worked 25 hours last week and 15 hours this week. How many more hours did Kathy work last week than this week?

____

13 **HOBBIES** Michael had 120 stamps in his stamp collection. He bought 30 more stamps. How many stamps does Michael have in his stamp collection now?

____

14 **BOOKS** Cara read 35 pages Monday and 32 pages Tuesday. How many pages did Cara read in the two days combined?

____

15 **NUTRITION** A blueberry granola bar has 98 Calories. An oatmeal granola bar has 110 Calories. How many more Calories does the oatmeal bar have than the blueberry bar?

____

16 **SOFTBALL** The Cardinals have 14 players on their team. Nine players are on the field when the team is playing defense. How many players are *not* on the field when the team is playing defense?

____

17 **MOVIES** The lower section of the theater has 60 seats. The upper section has 85 seats. How many total seats are in the theater?

____

Solve.

18 **SOCCER** Antonio scored 13 goals last soccer season. He scored 20 goals this soccer season. How many more goals did Antonio score this soccer season than last soccer season?

____

19 **FITNESS** Rena recorded her jogging times. How many minutes did Rena jog on Monday and Tuesday combined?

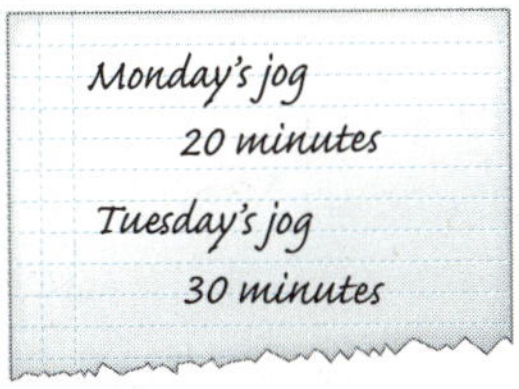

____

20 **ART** Mitch drew 5 pictures in art class during the first quarter. He drew 6 pictures during the second quarter. How many pictures did Mitch draw in art class during the first and second quarters?

____

21 **COLLECTIONS** Grace has 50 holiday snow globes in her collection. She also has 23 other snow globes. How many snow globes does Grace have in all?

____

GO ON

Vocabulary Check **Write the vocabulary word that completes each sentence.**

22 The answer to a subtraction problem is called the ______________________.

23 The answer to an addition problem is called the ______________________.

24 **Writing in Math** Sallie is 53 inches tall. Dulce is 61 inches tall. How much taller is Dulce than Sallie? Answer the question. Then describe how you know which operation to use to solve this problem.

____

____

Name the operations needed to solve each problem. Write one or more number sentences to solve each problem. Answer the question.

25 **HOBBIES** Matt is building a kite. He needs four pieces of wood. The pieces need to be 30 inches, 20 inches, 15 inches, and 10 inches in length. How many inches of wood does Matt need to build the kite?

____

26 **FOUR-WHEELING** Anna plans to go four-wheeling with friends. She wants to travel the shortest path. How many more kilometers is the Sand than the Forest trail?

____

27 **MUSIC** Yancy spent $5 downloading music in February, $32 downloading music in March, and $13 downloading music in April. How much more money did Yancy spend downloading music in March than in February and April?

____

____

28 **SALES** A store sold 33 pairs of shoes before noon. Before the store closed, 4 of these pairs were returned, and 52 more pairs were sold. At the end of the day, how many shoes were sold and not returned?

____

____

STOP

Lesson 1-2

Multiplication and Division Operations

3AF1.0 Students select appropriate symbols, operations, and properties to represent, describe, simplify, and solve simple number relationships.

KEY Concept

Operation	Answer	Key Words	Symbol
multiplication	product	times, each, per	× or •
division	quotient	divide, each, separate, per	÷ or $\frac{a}{b}$ or $b\overline{)a}$

multiplication (2 × 7)

$2 \times 7 = 14$ ← product

$8 \div 4 = 2$ ← quotient

division (8 ÷ 4)

VOCABULARY

multiplication
an operation on two numbers to find their *product*

product
the answer to a *multiplication* problem

division
an operation on two numbers in which the first number is separated into the same number of equal groups as the second number

quotient
the answer to a *division* problem

Multiplication is repeated addition. When used with *each* and *per*, the words *total*, *in all*, and *combined* can indicate multiplication. Division problems usually ask for the number of objects in each group or the number of equal groups.

Example 1

GARDENING Mr. Fernandez's class is planting a rose garden. They planted 4 rows of bushes. Each row had 8 bushes. How many bushes did the class plant in all?

1. You are asked how many rose bushes in all.
2. Write a multiplication sentence.

rows of bushes	×	bushes in each row	=	total number of bushes
4	×	8	=	32

3. Answer the question.
The class planted 32 rose bushes.

GO ON

YOUR TURN!

SPORTS Three times as many people watched the championship game as watched the last regular season game. Four thousand people watched the last regular season game. How many people watched the championship game?

1. You are asked ______________ people watched the championship game.
2. Write a ______________ sentence.

3 times	×	number who watched the last game	=	number who watched the championship
______	×	________	=	________

3. Answer the question. ______________________________

Example 2

BAKING Ms. Sanchez passed out 12 cookies equally to 3 students at an after-school club. How many cookies did each student get?

1. You are asked how many cookies each student will get.
2. Write a division sentence.

total number of cookies	÷	number of sudents sharing cookies	=	number of cookies for each student
12	÷	3	=	4

3. Answer the question.
   Each student gets 4 cookies.

YOUR TURN!

MONEY Beth is saving $5 each month to buy a $40 MP3 player. How many months will it take her to save enough money?

1. You are asked ______________ it will take her to save enough money to buy the MP3 player.
2. Write a ______________ sentence.

total number of dollars	÷	number of dollars saved each month	=	number of months
______	÷	______	=	______

3. Answer the question. ______________________________

Who is Correct?

Name the operation needed to solve the problem. Write a number sentence to solve the problem. Answer the question.

SCHOOL SUPPLIES Lena bought 3 packs of markers. Each pack contained 8 markers. How many markers did Lena buy in all?

Dillon	Silvia	Cedro
multiplication;	**addition;**	**division;**
$3 \times 8 = 24$	$8 + 8 + 8 = 24$	$8 \div 3 = 2.67$

Circle correct answer(s). Cross out incorrect answer(s).

Guided Practice

Name each operation modeled.

1

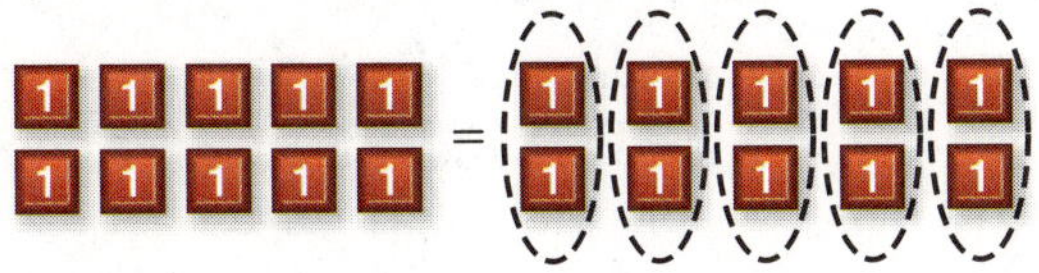

____

2

____

Step by Step Practice

Solve.

3 **SCHOOL** A school auditorium has 60 rows of seats. Each row has 10 seats. What is the total number of seats in the auditorium?

Step 1 You are asked for the ____________ number of seats in the auditorium. You know the number of seats in ____________ row.

Step 2 Write a ____________ sentence.
60 ______ 10 = ______

Step 3 Answer the question.
There are ______ seats in the auditorium.

GO ON

Name the operation needed to solve each problem. Write a number sentence to solve the problem. Answer the question.

4. **PACKAGING** The volleyball team ordered 27 new jerseys. There were 3 jerseys in each package. How many packages of jerseys did the volleyball team order?

____

5. **LANDSCAPING** Mrs. Bulach's class planted 5 rows of trees. Each row had 3 trees. How many trees did Mrs. Bulach's class plant in all?

____

6. **PHOTOS** Wang has a photo album with 20 pages. Each page holds 4 photos. How many photos can Wang put in the album in all?

____

Step by Step Problem-Solving Practice

Problem-Solving Strategies

- ☑ Use a model.
- ☐ Look for a pattern.
- ☐ Act it out.
- ☐ Solve a simpler problem.
- ☐ Work backward.

Solve.

7. **PART-TIME JOB** Troy earned $15 for walking a friend's dog. He mowed 3 lawns on Saturday. He earned $12 for each lawn. How much did Troy earn in all?

Understand Read the problem. Write what you know.

Troy earned ______ walking a friend's dog.

Troy mowed ______ lawns. Troy was paid ______ for each lawn.

Plan Pick a strategy. One strategy is to use a model. Use money to show $15. Use money to show $12. Make 3 stacks of $12.

Solve Count the money.

$12 × ______ = $______

______ + ______ = ______

money from walking dog + money from mowing lawns = total earned

Answer the question. Troy earned $______.

Check Make a model using cubes. Count 15 cubes. Then make 3 rows of 12 cubes. Count the cubes.

8 **HOBBIES** Linda buys 9 marbles and receives 3 more marbles for free. She separates the marbles evenly among her 3 younger sisters. How many marbles did each sister get? ________ Check off each step.

_____ **Understand**

_____ **Plan**

_____ **Solve**

_____ **Check**

9 **CLUBS** Mr. Devono wants to put 2 pencils on each desk in his room for a science club meeting. His classroom has 4 rows of desks. Each row has 6 desks. How many pencils does Mr. Devono need?

____

10 **TASTE TEST** The Smoothie Company wants to test new flavors for smoothies. They need 5 equal groups of people. A total of 41 people sign up for the taste test. On the day of the test, 6 people do not show up. How many people will be in each group?

____

11 **Reflect** How does drawing a model help with multiplication problems?

____

____

Skills, Concepts, and Problem Solving

Name each operation modeled.

12 

____

13 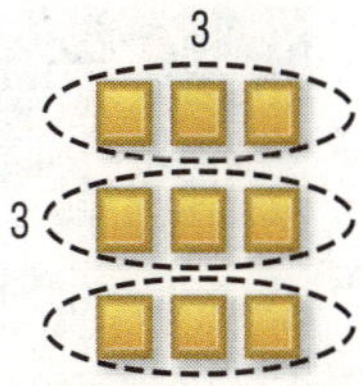

____

GO ON

Name the operation needed to solve each problem. Write a number sentence to solve the problem. Answer the question.

14 **PACKAGING** Gloria bought 8 packs of greeting cards. She spent a total of $32. How much did each pack of cards cost?

____

15 **COINS** James has 3 boxes of coins. Each box has 6 coins. How many coins does James have in all?

____

16 **PACKAGING** Mrs. Gomez bought 4 packages of hot dogs for a family reunion cookout. Each package had 8 hot dogs. How many hot dogs did Mrs. Gomez have in all?

____

17 **FITNESS** Randall ran around the track for 20 minutes. Each lap took him 2 minutes. How many laps did he run?

____

18 **LANDSCAPING** Mr. Rhodes planted 3 rows of flowers in his flower garden. Each row had 8 flowers. How many flowers did Mr. Rhodes plant in all?

____

Solve.

19 **COMIC BOOKS** Mrs. Patel has 20 comic books. She wants to give each of her 4 children 7 comic books. How many more comic books does she need?

____

20 **HOBBIES** McKenzie has sheets of stamps like the one shown at right. She has 5 sheets of stamps. She uses 11 stamps to mail packages and letters. How many stamps does she have left?

____

21 **SHOPPING** Ken has a coupon for $5 off his total purchase at a sports apparel store. He decides to buy 3 jerseys that cost $14 each. After he uses his coupon, how much will he pay for the jerseys?

____

HOBBIES Sheets of stamps

Vocabulary Check **Write the vocabulary word that completes each sentence.**

22 An operation on two numbers to find their product is called

_______________.

23 The answer to a division problem is called its _______________.

24 **Writing in Math** Rosa made 8 beaded bracelets. She used a total of 48 beads to make the bracelets. How many beads did Rosa use to make each bracelet if each bracelet had the same number of beads? Solve. Explain how you can solve this problem using a model.

____

____

Spiral Review

Name the operation needed to solve each problem. Write a number sentence to solve the problem. Answer the questions. (Lesson 1-1, p. 4)

25 **FOOD** Connie ordered 6 hot dogs and 5 hamburgers. How many food items did she order in all?

____

26 **SCHOOL** Review Jaul's scores on his science exams. How many more points did he score on his second exam than his first exam?

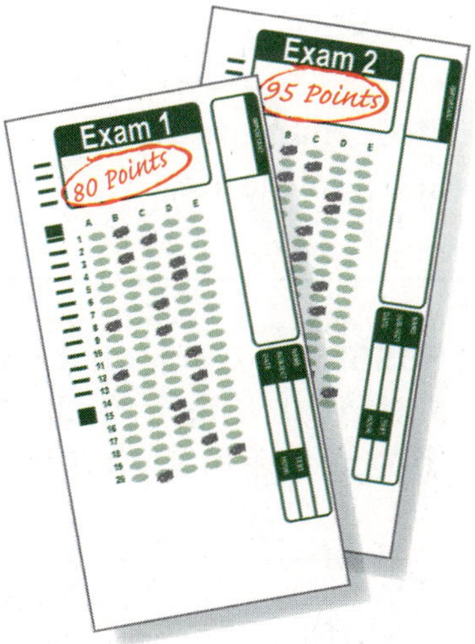

____

Solve. (Lesson 1-1, p. 4)

27 **PHOTOS** Lesley took 25 pictures on Friday. She took 38 pictures on Saturday. How many more pictures did Lesley take on Saturday than on Friday?

____

28 **WEATHER** In January it snowed 23 inches, and in February it snowed 17 inches. How many inches did it snow in January and February combined?

____

STOP

Chapter 1 Progress Check 1 (Lessons 1-1 and 1-2)

Name each operation modeled.

1 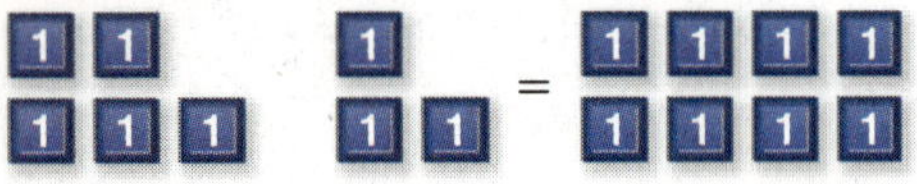

____

2 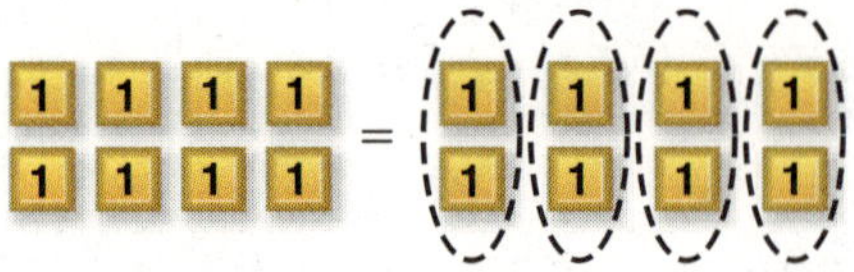

____

Name the operation needed to solve each problem. Write a number sentence to solve the problem. Answer the question.

3 **SPORTS** The track coach received 8 boxes of spikes. Each box contained 12 spikes. How many spikes did he receive in all?

____

4 **TEMPERATURE** It was 92°F on Saturday and 85°F on Sunday. How much warmer was it on Saturday than Sunday?

____

5 **READING** Marissa read 230 pages in 5 days. She read the same number of pages each day. How many pages did she read each day?

____

Solve.

6 **TRAVEL** The map shows the distances that the Cooper family drove on Tuesday and Wednesday. How many miles did they drive in all on Tuesday and Wednesday?

____

7 **FOSSILS** A scientist found 92 fossils in 2004 and 200 fossils in 2005. How many more fossils did the scientist find in 2005 than in 2004?

____

8 **SCHOOL** For homework, Kyle was assigned 16 problems from one lesson and 7 problems from another. How many problems does Kyle have for homework in all?

____

Lesson 1-3

Equality

4AF2.1 Know and understand that equals added to equals are equal.

4AF2.2 Know and understand that equals multiplied by equals are equal.

KEY Concept

An **equation** is a mathematical sentence that contains an equal sign. It is like a balance scale that is level.

To keep the scale level, what you do to one side of the equation, you must do to the other side.

Property	Definition	Example
Addition Property of Equality	**If you add the same number to each side of an equation, the two sides remain equal.**	$6 = 6$ $6 + 2 = 6 + 2$ $8 = 8$
Multiplication Property of Equality	**If you multiply by the same number on each side of an equation, the two sides remain equal.**	$5 = 5$ $5 \cdot 2 = 5 \cdot 2$ $10 = 10$

These properties are more commonly used with addition and multiplication, but also apply to subtraction and division.

VOCABULARY

equation
a mathematical sentence that contains an equal sign, =, indicating that the amount on the left side of the equal sign has the same value as the amount on the right
example: $2 \times 5 = 6 + 4$

equal
having the same value

Addition Property of Equality
adding the same amount to each side of an equation results in a true equation

Multiplication Property of Equality
multiplying each side of an equation by the same amount results in a true equation

Example 1

Show that adding 2 to each side of $4 + 3 = 6 + 1$ results in a true equation.

1. Make a model showing $4 + 3 = 6 + 1$.

   This equation is balanced because $4 + 3 = 7$ and $6 + 1 = 7$.

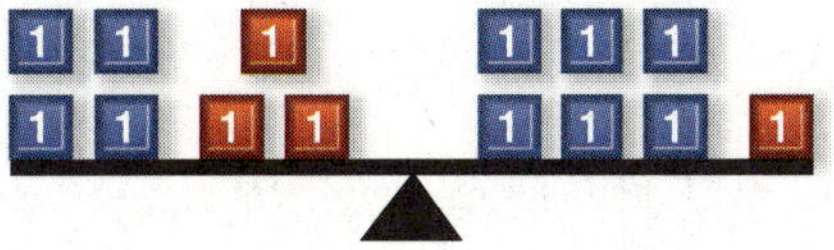

2. Add 2 to each side of the model.

   This equation is balanced or true because $(4 + 3) + 2 = 7 + 2$ or 9 and $(6 + 1) + 2 = 7 + 2$ or 9.

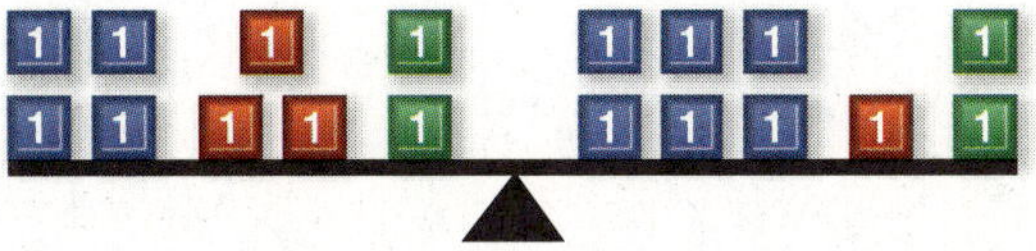

GO ON

YOUR TURN!

Show that multiplying each side of $3 + 1 = 2 + 2$ by 2 results in a true equation.

1. Make a model showing $3 + 1 = 2 + 2$. Why is this equation balanced?

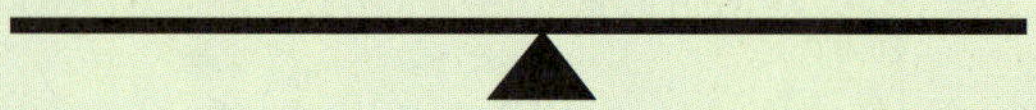

____

2. Double each side of the model to show multiplication by 2. Why is this equation balanced?

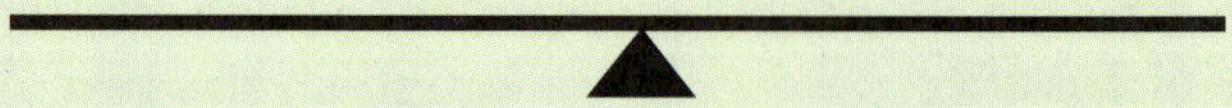

____

____

Example 2

What number goes in the blank to make $(7 - 5) \times 4 = 2 \times$ ______ a true equation?

1. The expressions $7 - 5$ and 2 have the same value.

   $7 - 5 = 2.$

   Both sides of the equation involve multiplication.

2. Identify and apply the correct property to find the missing number.

   Use the Multiplication Property of Equality.

   $(7 - 5) \times 4 = 2 \times 4$

3. Check by showing that $(7 - 5) \times 4 = 2 \times 4$ is a true equation.

   This equation is true because:

   $(7 - 5) \times 4 = 2 \times 4$

   $2 \times 4 = 2 \times 4$

   $8 = 8$ ✔

YOUR TURN!

What number goes in the blank to make $3 + (18 \div 2) =$ ______ $+ (14 - 5)$ a true equation?

1. The expressions __________ and __________ have the same value.

   __________ = __________.

   Both sides of the equation involve __________.

2. Identify and apply the correct property to find the missing number.

   Use the __________ Property of Equality.

   $3 + (18 \div 2) =$ ______ $+ (14 - 5)$

3. Check by showing that $3 + (18 \div 2) =$ ______ $+ (14 - 5)$ is a true equation.

   This equation is true because:

   $3 + (18 \div 2) =$ ______ $+ (14 - 5)$

   $3 + 9 =$ ______ $+ 9$

   ______ = ______ ✔

Who is Correct?

Show that adding 1 to each side of 7 + 9 = 8 + 8 results in a true equation.

Hao

$(7 + 9) \cdot 1 = (8 + 8) \cdot 1$
$16 = 16$

Charlotte

$(7 + 9) + 1 = (8 + 8) + 1$
$16 + 1 = 16 + 1$
$17 = 17$

Marisela

$7 + 9 + 1 = 8 + 8 + 1$
$7 + 10 = 8 + 9$
$17 = 17$

Circle correct answer(s). Cross out incorrect answer(s).

Guided Practice

1 Show that multiplying by 2 on each side of $2 \cdot 3 = 6 \cdot 1$ results in a true equation.

____

____

2 Show that adding 5 to each side of $7 + 2 = 6 + 3$ results in a true equation.

____

____

Step by Step Practice

3 What number goes in the blank to make $(8 + 5) + 6 =$ ______ $+ (20 - 7)$ a true equation?

Step 1 The expressions inside parentheses have the same value.
$8 + 5 =$ __________.

Step 2 Identify and apply the correct property to find the missing number.

Use the __________ Property of Equality.

$(8 + 5) + 6 =$ ______ $+ (20 - 7)$

Step 3 Check: $(8 + 5) + 6 =$ ______ $+ (20 - 7)$
$13 + 6 =$ ______ + ______
______ = ______

> The left side of the equal sign has the same value as the right side.

GO ON

Find the missing number to make each equation true.

4 _____(18) = (16 + 2)5

Since _____ = ________,

_____(18) = (16 + 2)5

by the _______________

Property of Equality.

5 1 + (20 ÷ 5) = (9 − 5) + _____

Since ________ = ________,

1 + (20 ÷ 5) = 9 − 5 + _____

by the _______________

Property of Equality.

Step by Step Problem-Solving Practice

Problem-Solving Strategies

- ☐ Look for a pattern.
- ☐ Guess and check.
- ☐ Act it out.
- ☑ Solve a simpler problem.
- ☐ Work backward.

Solve.

6 **MONEY** Ronni had 10 dollars. She earned 3 more dollars. Nick had 13 dollars. Then Ronni and Nick earned 4 dollars each. Do Ronni and Nick have the same amount of money now?

Understand Read the problem. Write what you know.

Ronni has _____ + _____ dollars.

Nick has _____ dollars.

Each earned _____ dollars more.

Plan Pick a strategy. One strategy is to solve a simpler problem. Find the total dollars for each.

Solve Solve the equation for Ronni's money.

____

Solve the equation for Nick's money.

____

Do Ronni and Nick have equal amounts of money?

____

Written as one equation:

Ronni		Nick
(10 + 3) + 4	=	13 + 4

Check You can make a model to check your work.

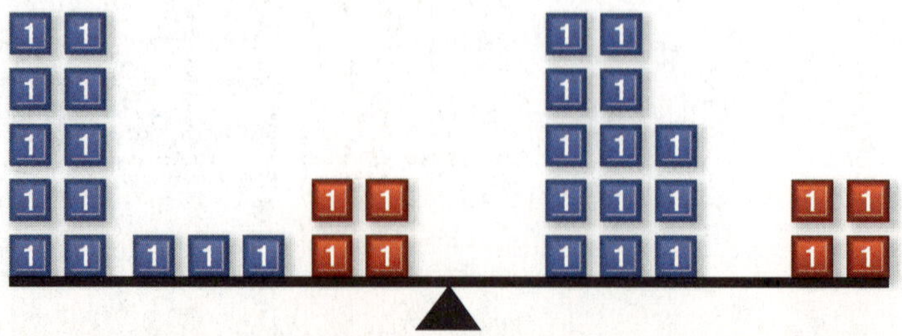

7 **HOBBIES** Nora had 12 dolls. She bought 2 more. Mattie had 9 dolls. She bought 5 more. Nora and Mattie each received 2 dolls for their birthdays. How many dolls do Nora and Mattie each have? Check off each step.

____

_____ **Understand**

_____ **Plan**

_____ **Solve**

_____ **Check**

8 **SHOPPING** Jeans are on sale for \$24 each. Shirts are on sale for \$11 each. Kaya has a coupon for \$5 off any purchase. She wants to buy 2 shirts. Will she spend the same amount if she buys 1 pair of jeans? If not, which will cost less?

____

9 **Reflect** How are the Addition and Multiplication Properties of Equality the same?

____

____

Skills, Concepts, and Problem Solving

10 Show that adding 7 on each side of $1 + 3 = 2 + 2$ results in a true equation.

____

____

11 Show that multiplying by 2 on each side of $3 \cdot 4 = 2 \cdot 6$ results in a true equation.

____

____

Find the missing number to make each equation true.

12 $(15 \cdot 2) + \_\_\_\_\_ = 10 + (6 \cdot 5)$

13 $4 \cdot (19 - 1) = (13 + 5) \cdot \_\_\_\_\_$

14 $5 + (24 \div 4) = 5 + (2 \cdot \_\_\_\_\_)$

15 $8 \cdot (3 + \_\_\_\_\_) = 8 \cdot (12 - 7)$

GO ON

Solve.

16 **WEATHER** The graph shows the amount of snowfall in Colorado and Michigan over three days. After Wednesday, did Colorado and Michigan have the same amount of snow? Explain. ______________

____

____

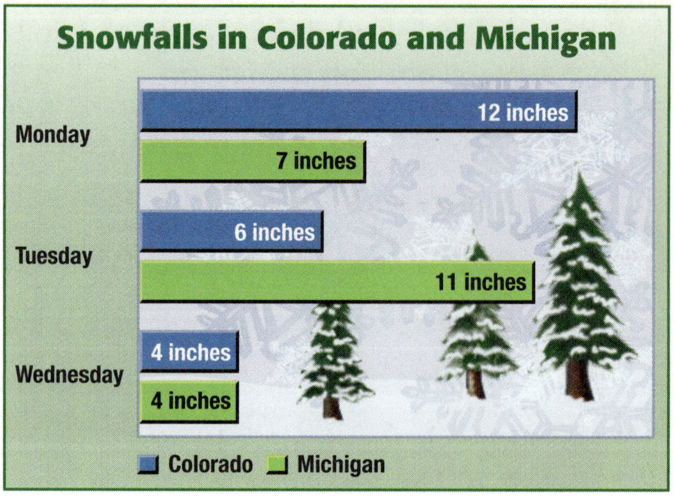

17 **MOVIES** Cameron and Yoko went to see a movie. Cameron spent $3 on a soda, $6 on popcorn, and $2 on candy. Yoko spent $4 on a soda and $7 on candy. If Cameron and Yoko paid $10 for each of their tickets, did they spend the same amount of money at the movies? Explain.

____

Vocabulary Check **Write the vocabulary word that completes each sentence.**

18 The ______________ Property of Equality states that multiplying each side of an equation by the same amount results in a true equation.

19 **Writing in Math** Explain the meaning of the equal sign (=).

____

____

Spiral Review

Name the operation needed to solve the problem. Write a number sentence to solve the problem. Answer the question.

20 **ADVERTISING** Mrs. Rodriguez paid for 3 newspaper ads. Each ad ran for the same number of days. How many days did each ad appear if she was charged for a total of 30 days? (Lesson 1-2, p. 11)

____

21 **POPULATION** Last year 540 people lived in Nelsonville. This year 610 people live there. How many more people live in Nelsonville this year than last year? (Lesson 1-1, p. 4)

____

STOP

Lesson 1-4

Operations with Unknown Quantities

KEY Concept

You can use a letter, a box, or other symbols to represent an unknown amount or quantity. These symbols are called **variables**.

Expressions	*Equations*
$5 + \square$	$10 \cdot \square = 20$
$8 - a$	$16 \div d = 4$

Inverse operations are opposite operations. They undo each other. Addition and subtraction are inverse operations. Multiplication and division are also inverse operations.

Inverse Operations	Definition	Example
addition and subtraction	When $a + b = c$, then $c - a = b$ and $c - b = a$	$7 + 8 = 15$, so $15 - 7 = 8$ and $15 - 8 = 7$
multiplication and division	When $a \cdot b = c$, then $c \div a = b$ and $c \div b = a$	$4 \cdot 3 = 12$, so $12 \div 4 = 3$ and $12 \div 3 = 4$

4AF1.1 Use letters, boxes, or other symbols to stand for any number in simple expressions or equations.

VOCABULARY

variable
a letter or symbol used to represent an unknown quantity
Example: $5 + x = 10$
↳ variable (x)

inverse operations
operations that undo each other

To undo addition, use subtraction. To undo subtraction, use addition. To undo multiplication, use division. To undo division, use multiplication.

Example 1

What number belongs in the $\square$ to make the equation $6 + \square = 9$ true?

1. Use the fact that addition and subtraction are inverse operations.

   $6 + \square = 9$, so $9 - 6 = \square$ $\quad$ $9 - 6 = 3$, so $\square = 3$

2. Use a model to check your answer.
   Think: What number added to 6 equals 9?
   $6 + 3 = 9$ $\quad$ The value of $\square$ must be 3.

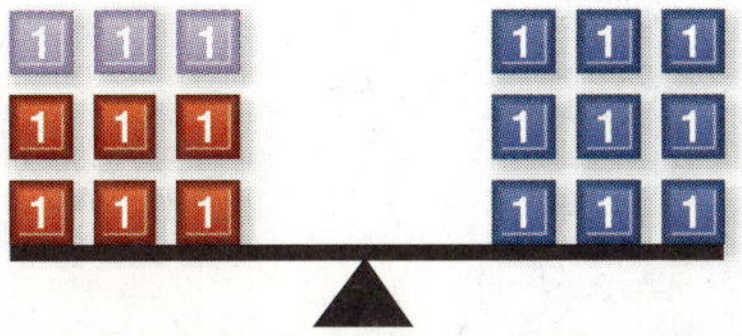

GO ON

YOUR TURN!

What number belongs in the □ to make $7 + \square = 18$ true?

1. Use the fact that addition and ____________ are inverse operations.
2. Use a model to check your answer.
   Think: What number added to 7 equals 18?

   $7 + \square = 18$

   The value of □ must be __________.

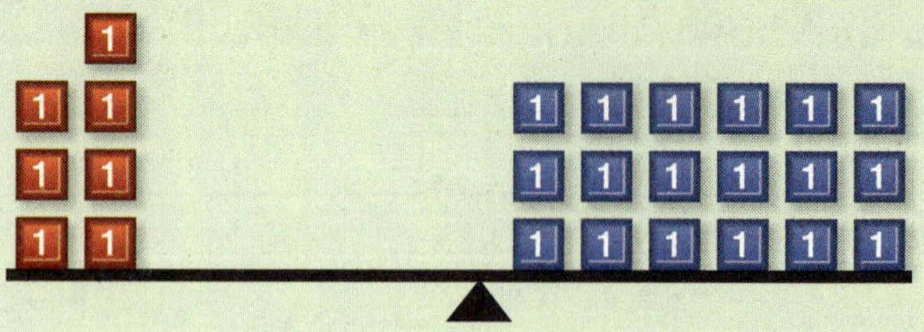

Example 2

Find the value of c in the equation $3 \cdot c = 18$.

1. Use the fact that multiplication and division are inverse operations.
   $3 \cdot c = 18$, so $c = 18 \div 3$
   $18 \div 3 = 6$, so $6 = c$
2. Check your answer by substituting 6 for c.
   $3 \cdot c = 18$
   $3 \cdot 6 = 18$
   $18 = 18$ ✔

YOUR TURN!

Find the value of y in the equation $8 \cdot y = 72$.

1. Use the fact that multiplication and division are inverse operations.
   $8 \cdot y = 72$, so ______ $= 72 \div 8$
   $72 \div 8 =$ ______, so ______ $= y$
2. Check your answer by substituting ______ for y.
   $8 \cdot y = 72$
   $8 \cdot$ ______ $= 72$
   ______ $= 72$

Who is Correct?

Find the value of the variable in the equation $3 \cdot t = 96$.

Dora
$3 \cdot t = 96$
$96 - 3 = t$,
so $t = 93$

Rose
$3 \cdot t = 96$, so
$96 \cdot 3 = t$
$t = 288$

Neal
$3 \cdot t = 96$, so
$t = 96 \div 3$
$32 = t$

Circle correct answer(s). Cross out incorrect answer(s).

Guided Practice

Find the value of each box or variable by modeling the equation.

1 $4 + \square = 10$

$\square =$ ______

2 $11 - b = 7$

$b =$ ______

Step by Step Practice

3 Find the value of z in the equation $\frac{z}{8} = 4$.

Step 1 $\frac{z}{8}$ means $z \div 8$. Use the fact that multiplication and division are inverse operations.

$z \div 8 =$ ______, so $z = 8 \cdot$ ______

$8 \cdot$ ______ $=$ ______, so $z =$ ______

Step 2 Check your answer by substituting ______ for z.

$\frac{z}{8} = 4$

$\frac{\square}{8} = 4$

______ $= 4$ ✔

Find the value for the box or the variable in each equation.

4 $\square - 17 = 9$

$\square - 17 = 9$, so $\square = 9 +$ ______

$9 +$ ______ $=$ ______, so ______ $= \square$

5 $4 \times n = 28$

$4 \times n = 28$, so $n = 28 \div$ ______

$28 \div$ ______ $=$ ______, so ______ $= n$

6 $\frac{y}{3} = 9$

$y =$ ______

7 $13 + x = 21$

$x =$ ______

GO ON

Step by Step Problem-Solving Practice

Problem-Solving Strategies

- ☐ Draw a diagram.
- ☐ Guess and check.
- ☐ Use a model.
- ☐ Solve a simpler problem.
- ☑ Write an equation.

Solve.

8 **SNACKS** Mr. Fox brought 32 oranges to a class party. There were 7 oranges left after the party. How many oranges were eaten during the party?

Understand Read the problem. Write what you know.

Mr. Fox brought ______ oranges.

There were ______ oranges left.

The key word left means to ________.

Plan Pick a strategy. One strategy is to write an equation. Then solve the equation.

Solve Let a represent the number of oranges eaten. Write an equation. Solve the equation.

Number of oranges brought to the party	−	Number of oranges eaten during the party	=	Number of oranges left after the party
32	−	a	=	7

If $32 - a = 7$, then $a + 7 = 32$

$a + 7 = 32$

Use the inverse operation of addition, which is subtraction.

$a + 7 -$ ______ $= 32 -$ ______

$a =$ ______

There were ______ oranges eaten during the party.

Check Substitute ______ for a.

$32 - a = 7$

$32 -$ ______ $= 7$

$7 = 7$ ✔

Write an equation to represent each situation, then answer the question.

9 **ENTERTAINMENT** Lola and Steve went to the ball game. They bought snacks that cost \$7.50. The total cost of the game tickets and snacks was \$23.50. How much did each game ticket cost? Check off each step.

____

_______ **Understand**

_______ **Plan**

_______ **Solve**

_______ **Check**

10 **FOOD** Marco is packaging doughnuts to sell at the fair. He is using bags that hold 12 doughnuts each. How many of these bags will he need to package 192 doughnuts?

____

11 **ELECTIONS** Mrs. Davis was running for school board. She had 225 campaign buttons to hand out. After one week, she had 36 buttons left. How many buttons did she hand out that week?

____

12 **Reflect** How do you decide which operation to perform to solve an equation that contains a variable?

____

____

Skills, Concepts, and Problem Solving

Find the value of each box or variable by modeling the equation.

13 $2m = 14$

$m =$ _______

14 $6 + \square = 13$

$\square =$ _______

GO ON

Find the value of the box or the variable in each equation.

15 $2n = 16$

$n =$ ______

16 $\square - 12 = 8$

$\square =$ ______

17 $19 + \square = 27$

$\square =$ ______

18 $\frac{t}{4} = 5$

$t =$ ______

Write an equation to represent each situation, then answer the question.

19 **FUND-RAISING** Johnny is a member of Mr. Alvarez's class. Johnny sold 52 items from the school's fund-raising catalog. How many items did the rest of his class sell?

Fund-Raiser Class Totals	
Class Teacher	Items Sold
Mr. Alvarez	176
Ms. Williams	205
Ms. Patterson	145

____

20 **JOBS** José earned \$13 per hour last week. His total earnings were \$325. How many hours did José work last week?

____

Vocabulary Check **Write the vocabulary word or term that completes each sentence.**

21 A ______________ is a letter or symbol used to represent an unknown quantity.

22 **Writing in Math** Explain why multiplication and division are inverse operations. Include an example.

____

____

Spiral Review

23 Show that adding 3 to each side of $18 + 5 = 25 - 2$ results in a true equation. (Lesson 1-3, p. 19)

____

____

____

STOP

Chapter 1

Progress Check 2 (Lessons 1-3 and 1-4)

1 Show that adding 6 to each side of $1 + 4 = 2 + 3$ results in a true equation.

______

______

______

2 Show that multiplying by 2 on each side of $1 \cdot 4 = 2 \cdot 2$ results in a true equation.

______

______

______

Find the value for the box or the value of each variable by modeling the equation.

3 $9 - n = 4$ $n =$ ______

4 $2 + \square = 10$ $\square =$ ______

5 What number goes in the blank to make $(30 \div 2) + 4 = 15 +$ ______ a true equation?

______

6 What number goes in the blank to make ______ $\cdot (8 - 1) = 20(14 \div 2)$ a true equation?

______

Find the value for the box or the variable in each equation.

7 $\frac{m}{4} = 7$ $m =$ ______

8 $15 + \square = 34$ $\square =$ ______

9 $50 - \square = 13$ $\square =$ ______

10 $9x = 108$ $x =$ ______

Write an equation to represent each situation. Then answer the question.

11 **MONEY** Jade earned $10 per hour last week. Her total earnings were $250. How many hours did Jade work last week?

______

12 **AGES** Marcus is 15 years old. The difference between his age and his younger brother's age is 3 years. How old is his younger brother?

______

Chapter 1

Study Guide

Vocabulary and Concept Check

Addition Property of Equality, *p. 19*
difference, *p. 4*
inverse operations, *p. 25*
Multiplication Property of Equality, *p. 19*
product, *p. 11*
quotient, *p. 11*
sum, *p. 4*
variable, *p. 25*

Write the vocabulary word that completes each sentence.

1. A(n) ________ is the answer to a *subtraction* problem.
2. Addition and subtraction are ____________.
3. If $9 = 9$, then $9 \cdot 9 = 9 \cdot 9$ represents the ______________________.

Write the correct vocabulary term in each blank.

4. ____________ (points to 3) $9 \div 3 = 3$

5. ____________ (points to 40) $5 \times 8 = 40$

6. ____________ (points to 11) $6 + 5 = 11$

Lesson Review

1-1 Addition and Subtraction Operations (pp. 4–10)

Name the operation needed to solve each problem. Write a number sentence to solve each problem. Answer the question.

7. **FARMING** Mr. Kitts planted 12 acres of soybeans. His neighbor planted 37 acres of soybeans. How many acres did they plant in all?

____

8. **DESIGN** Madison wants to purchase a new bedspread. She has two choices. One bedspread costs \$88, and the other costs \$47. What is the difference in their prices?

____

Example 1

BAKE SALE **Ahmed took 12 cookies to the bake sale. Hector took 16 cookies. How many more did Hector take than Ahmed?**

You are asked how many more cookies Hector took than Ahmed.

Write a subtraction sentence.

Hector's cookies	–	Ahmed's cookies	=	difference in number of cookies
16	–	12	=	4

Hector took 4 more cookies than Ahmed.

1-2 Multiplication and Division Operations (pp. 11–17)

Name the operation needed to solve each problem. Write a number sentence to solve the problem. Answer the question.

9. **PACKAGING** The school band ordered 5 cases of concert band shirts. If there are 12 shirts per case, then what was the total number of shirts they ordered?

____

____

10. **ART** Tia bought 4 prints at the art show for $6 each. What was the total bill for the art she bought?

____

____

11. **NUTRITION** A mini-bagel box contains 3 bagels. The total number of Calories in the entire box of bagels is 360. How many Calories are in each bagel?

____

____

12. **SHOPPING** The Taylor Department Store has 15 empty clothing racks. The racks must hold 180 dresses. If each rack is to have the same number of dresses, then how many dresses should be placed on each rack?

____

____

Example 2

CARDS Tabitha is playing a game. She must place 3 rows of 6 cards each facedown. How many total cards did Tabatha place facedown?

You are asked how many total cards Tabitha placed facedown.

Write a multiplication sentence.

number of rows	×	number of cards in each row	=	total number of cards
3	×	6	=	18

She placed a total of 18 cards facedown.

Example 3

FOOD Mr. Martin had 20 peanuts for his 4 children to share. How many peanuts did each child get?

You are asked how many peanuts each child received.

Write a division sentence.

total number of peanuts	÷	number of children	=	number of peanuts for each child
20	÷	4	=	5

Each child gets 5 peanuts.

1-3 Equality (pp. 19–24)

13 Show that adding 9 to each side of $5 + 3 = 2 + 6$ results in a true equation.

____

____

____

14 Show that multiplying by 2 on each side of $5 \cdot 9 = 3 \cdot 15$ results in a true equation.

____

____

____

Find the missing number to make each equation true.

15 $(10 + 2) + \_\_\_\_ = 18 + (4 + 8)$

16 $\_\_\_\_(9 \cdot 6) = (30 + 24) \cdot 13$

17 $12 - 6 + 6 = \_\_\_\_ + 5$

18 $2 \cdot 3 \cdot 8 = 2 \cdot 6 \cdot \_\_\_\_$

19 $(20 - \_\_\_\_) + 1 = (14 + 4) + 1$

20 **SONGS** Lela has 56 songs downloaded. She plans to download 20 more songs tomorrow. Ernesto has 55 songs downloaded and will download 20 tomorrow. Will they have the same number of songs downloaded after tomorrow?

____

Example 4

Show that adding 4 to each side of $3 + 2 = 4 + 1$ results in a true equation.

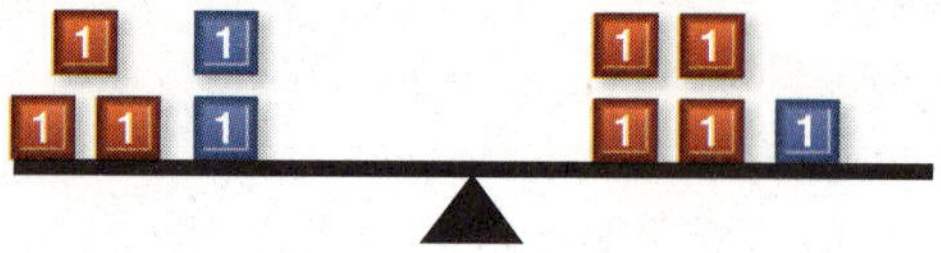

Make a model showing $3 + 2 = 4 + 1$.
This equation is balanced, or true, because $3 + 2 = 5$ and $4 + 1 = 5$.

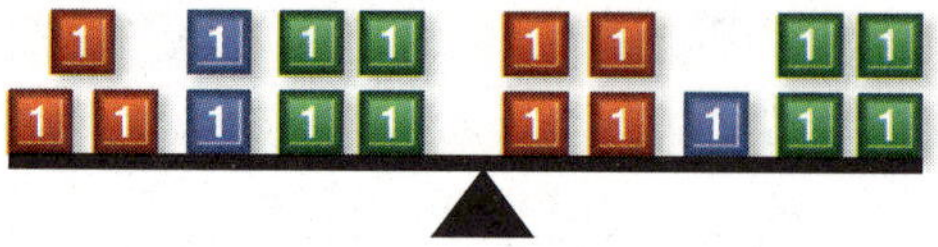

Add 4 to each side of the model.
This equation is balanced because
$(3 + 2) + 4 = 5 + 4 = 9$ and
$(4 + 1) + 4 = 5 + 4 = 9$.

Example 5

What number goes in the blank to make $3(18 - 3) = \_\_\_\_ \cdot (7 + 8)$?

Analyze the equation.

$18 - 3$ and $7 + 8$ both equal 15.

So $18 - 3 = 7 + 8$.

Identify and apply the correct property.

Since $18 - 3 = 7 + 8$, and $18 - 3$ is multiplied by 3, use the Multiplication Property of Equality.

$3(18 - 3) = 3(7 + 8)$

Check: $3(18 - 3) = 3(7 + 8)$

$3(15) = 3(15)$

$45 = 45$

1-4 Operations with Unknown Quantities (pp. 25–30)

Find the value of the variable in the equation.

21 $x + 11 = 19$

$x =$ ______

22 $b - 3 = 31$

$b =$ ______

23 $y + 10 = 50$

$y =$ ______

24 $m - 21 = 24$

$m =$ ______

Example 6

Find the value of a in the equation $a - 6 = 2$.

Use the fact that addition and subtraction are inverse operations.

$a - 6 = 2$, so $a = 2 + 6$

$2 + 6 = 8$, so $8 = a$

Use a model to check your answer.
Think: What number minus 6 equals 2?

$8 - 6 = 2$

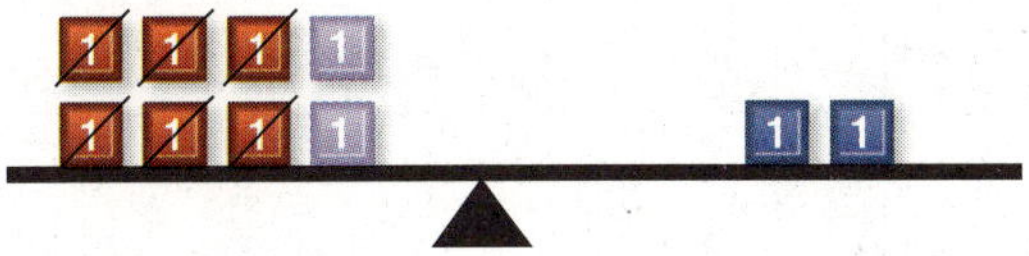

The value of a must be 8.

Find the value of the variable in the equation.

25 $\frac{y}{7} = 4$

$y =$ ______

26 $f \cdot 5 = 45$

$f =$ ______

27 $\frac{t}{2} = 12$

$t =$ ______

28 $b \cdot 8 = 56$

$b =$ ______

Example 7

Find the value of d in the equation $4 \cdot d = 24$.

Think: What number times 4 equals 24?

Use the fact that multiplication and division are inverse operations.

$4 \cdot d = 24$, so $d = 24 \div 4$

$24 \div 4 = 6$, so $6 = d$

Check your answer by substituting 6 for d.

$4 \cdot d = 24$

$4 \cdot 6 = 24$

$24 = 24$

$24 = 24$ is a true statement, so 6 is correct.

Chapter 1

Chapter Test

Answer each question below.

1 Name a key word or phrase for *addition* that helps you choose which operation to use.

____

Name the operation needed to solve each problem. Write a number sentence to solve the problem. Answer the question.

2 **HOMEWORK** Casey had a 10-page paper to write for language arts class. She was finished with 4 pages. How many did she have remaining?

____

3 **SHOPPING** Jason bought 7 DVDs that cost $14 each. How much did he pay in all?

____

4 **BAKING** Karen had a 16-inch roll of cookie dough that she cut into 2-inch segments. How many segments could she get out of her roll?

____

5 **JOBS** Ramon earned $30 walking dogs. He earned $5 more for babysitting his little brother. How much did he earn altogether?

____

6 Show that multiplying by 3 on each side of $2 \cdot 6 = 4 \cdot 3$ results in a true equation.

____

____

____

7 Show that adding 6 to each side of $15 + 10 = 5 + 20$ results in a true equation.

____

____

____

Find the missing numbers to make each equation true.

8 $(7 + 1) \cdot 16 =$ ______$(4 + 4)$

9 ______ $+ (12 \cdot 2) = (3 \cdot 8) + 12$

GO ON

Find the value of the variable in each equation.

10 $\square - 14 = 8$

$\square =$ ______

11 $9 \times z = 72$

$z =$ ______

12 $x + 71 = 82$

$x =$ ______

13 $\frac{\square}{9} = 6$

$\square =$ ______

Complete each sentence below.

14 The inverse operation of subtraction is ________.

15 The inverse operation of multiplication is ________.

Write an equation to represent each situation. Then answer the question.

16 **FIELD TRIPS** Carl's class took a field trip to a local dinner theater. There are 28 students in his class. The tickets cost $196 in all. What was the price of each ticket?

____

17 **BAKING** Denzel is making a pie. He has 12 cups of pecans. His mother brings him more pecans. He has 18 cups in all. How many cups did his mother bring him?

____

18 **SCIENCE** Mr. King has 90 test tubes. He must distribute 6 test tubes to each lab station in his classroom. How many lab stations in all?

____

19 **SCIENCE** The boiling point of Cobalt (Co) is 2,870°C. The boiling point of Iron (Fe) is 2,750°C. What is the difference in the two boiling points of these elements?

____

Correct the mistake.

20 **FOOD** Jackson worked at the concession stand during his high school's football game. One student bought $6 of snacks and paid Jackson with a $20 bill. Jackson gave him one $10 bill and one $5 bill as change. What mistake did Jackson make?

____

____

STOP

Chapter 1

Standards Practice

Choose the best answer and fill in the corresponding circle on the sheet at right.

1 **Which symbol makes the sentence true?**

$$45 \square 29 = 16$$

A $+$

B $\div$

C $\times$

D $-$

2 **Pao wrote $x \div 9 = 36$. Which sentence describes what he wrote?**

F 36 divided by 9 is x.

G The quotient of 36 and a number is 9.

H The quotient of a number and 9 is 36.

J 36 is the product of 9 and a number.

3 **A store is having a sale on soda. If Andrew only wants to buy one soda, how much will he pay?**

A $0.75

B $0.90

C $1.33

D $12.00

4 **Sasha's dad gave her $40 to take some friends to the movies. If movie tickets cost $8 per student, which equation will help Sasha figure out how many friends she can take? Let *n* equal the number of students going to the movies.**

F $\$8 - n = \$40$

G $\$8 \times n = \$40$

H $\$8 \div n = \$40$

J $\$40 - n = \$8$

5 **Laura and Evan baked 10 pies. Laura baked 3 pies. Which equation is used to find the number of pies Evan baked?**

A $3 + e = 10$

B $3 \times e = 10$

C $e - 3 = 10$

D $10 \div e = 3$

6 **Sonia can walk 4 miles an hour. How many miles can she walk in 3 hours?**

F 12 miles

G 9 miles

H 8 miles

J 7 miles

7 **Natalie is visiting her grandfather's horse farm. She is counting the number of legs she sees in the field. If she counts a total of 216 legs, how many horses are in the field?**

A 108 horses

B 72 horses

C 54 horses

D 27 horses

GO ON

8 Which operation will make the least answer?

45 □ 9 = ?

F +

H ×

G −

J ÷

9 Which operation is used with the term "product"?

A addition

C multiplication

B subtraction

D division

10 Mr. Rivera is making medals for the middle school's walk-a-thon participants. Each medal needs 18 inches of ribbon. If there are 54 participants in the walk-a-thon, how many inches of ribbon will he need altogether?

F 3 inches

H 72 inches

G 36 inches

J 972 inches

11 Find the next three numbers in this pattern.

896, 448, 224, 112, ____, ____, ____

A 94, 62, 31

C 60, 30, 15

B 28, 14, 7

D 56, 28, 14

12 Which symbol makes the sentence true?

718 □ 136 = 854

F +

H ×

G ÷

J −

ANSWER SHEET

Directions: Fill in the circle of each correct answer.

1 (A) (B) (C) (D)
2 (F) (G) (H) (J)
3 (A) (B) (C) (D)
4 (F) (G) (H) (J)
5 (A) (B) (C) (D)
6 (F) (G) (H) (J)
7 (A) (B) (C) (D)
8 (F) (G) (H) (J)
9 (A) (B) (C) (D)
10 (F) (G) (H) (J)
11 (A) (B) (C) (D)
12 (F) (G) (H) (J)

STOP

Math Fundamentals

Knowing math fundamentals, or the basic rules of math, can help you do things like save for a trip. For example, how much would it cost your family to fly to Utah if each plane ticket costs $120 and there are 4 people in your family?

STEP 1 Quiz

Are you ready for Chapter 2? Take the Online Readiness Quiz at ca.mathtriumphs.com to find out.

STEP 2 Preview

Get ready for Chapter 2. Review these skills and compare them with what you'll learn in this chapter.

What You Know

You know how to multiply.

Examples: $7 \times 4 = 28$
$4 \times 7 = 28$

TRY IT!

1. $5 \times 5 =$ ______
2. $10 \cdot 10 =$ ______
3. $7 \times 12 =$ ______
4. $2 \cdot 24 =$ ______

What You Will Learn

Lesson 2-1

The order in which you multiply numbers does not change the answer. This is called the **Commutative Property of Multiplication**.

$7 \cdot 4 = 28$
$4 \cdot 7 = 28$
The answer is the same.

What You Know

You know how to add.

Examples: $4 + 3 + 2 = 9$
$2 + 3 + 4 = 9$

TRY IT!

5. $3 + 5 + 10 =$ ______
6. $10 + 5 + 3 =$ ______
7. $20 + 3 + 3 =$ ______
8. $3 + 20 + 3 =$ ______

What You Will Learn

Lesson 2-2

Grouping numbers you are adding in different ways does not change the answer. This is called the **Associative Property of Addition**.

$(4 + 3) + 2 = 9$
$7 \quad + 2 = 9$

$4 + (3 + 2) = 9$
$4 + \quad 5 \quad = 9$

What You Know

You know how to add and multiply.

Example: $(2 \times 4) + (2 \times 5) = 18$
$8 \quad + \quad 10 \quad = 18$

TRY IT!

9. $(5 \cdot 2) + (3 \cdot 3) =$ ______
10. $(10 \cdot 9) + (1 \cdot 4) =$ ______
11. $(6 \times 3) + (4 \times 1) =$ ______
12. $(6 \cdot 3) + (4 \cdot 2) =$ ______

What You Will Learn

Lesson 2-3

Distribute a number by multiplying it with each member of a group. This is called the **Distributive Property**.

$$2(4 + 5) = (2 \times 4) + (2 \times 5)$$

Lesson 2-1

Commutative Property

2AF1.1 Use the commutative and associative rules to simplify mental calculations and to check results.
3AF1.5 Recognize and use the commutative and associative properties of multiplication.

KEY Concept

Property	Words	Definition	Example
Commutative Property of Addition	**The order in which two numbers are added does not change the sum.**	$a + b = b + a$	$3 + 5 = 5 + 3$ $8 = 8$
Commutative Property of Multiplication	**The order in which two numbers are multiplied does not change the product.**	$a \cdot b = b \cdot a$	$4 \cdot 5 = 5 \cdot 4$ $20 = 20$

VOCABULARY

Commutative Property of Addition
the order in which two numbers are added does not change the *sum*

Commutative Property of Multiplication
the order in which two numbers are multiplied does not change the *product*

sum
the answer to an addition problem (Lesson 1-1, p. 4)

product
the answer to a multiplication problem (Lesson 1-2, p. 11)

These properties tell you that order does not matter when you are adding or multiplying.

Example 1

Draw a model to show 2 + 4 = 4 + 2. Which property did you show?

1. Create a model for each side of the equation.

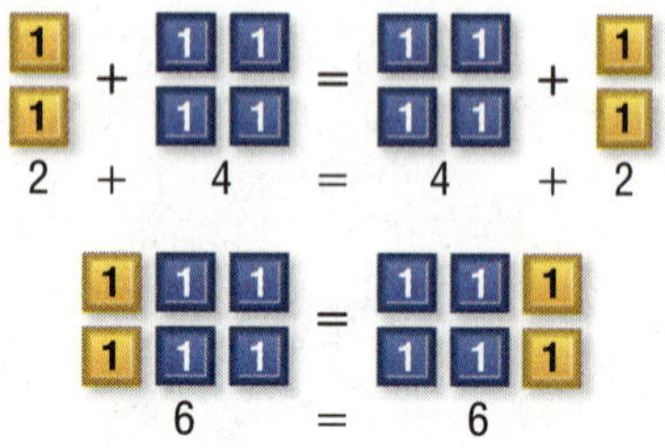

2. The order of the numbers changed, but the sum did not. This is the Commutative Property of Addition.

YOUR TURN!

Draw a model to show 3 • 2 = 2 • 3. Which property did you show?

1. Create a model for each side of the equation.

2. Which property did you show?

____

Example 2

Use the Commutative Property to fill in the blank. Check your answer.

$8 \cdot 6 =$ ______ $\cdot\ 8$

1. Use the Commutative Property of Multiplication.

   $8 \cdot 6 = \underline{\ \ 6\ \ } \cdot 8$

2. Check by multiplying the numbers on each side of the equation.

   $8 \cdot 6 = 6 \cdot 8$
   $48 = 48$

YOUR TURN!

Use the Commutative Property to fill in the blank. Check your answer.

$9 + 3 =$ ______ $+\ 9$

1. Use the Commutative Property of Addition.

   $9 + 3 =$ ______ $+ 9$

2. Check by adding the numbers on each side of the equation.

   ____

   ____

Who is Correct?

Give an example of the Commutative Property of Multiplication.

Ira	Cynthia	Diego
$7 \cdot 2 = 2 \cdot 7$ $14 = 14$	$7 \cdot 2 = 14$	$7 \cdot 2 = 14 \cdot 1$ $14 = 14$

Circle correct answer(s). Cross out incorrect answer(s).

Guided Practice

Draw a model to show each equation.

1 $2 + 5 = 5 + 2$
Which property did you show?

____

2 $3 \cdot 4 = 4 \cdot 3$
Which property did you show?

____

GO ON

Step by Step Practice

Use the Commutative Property to fill in the blank. Check your answer.

3 $12 + 9 = 9 + \_\_\_$

Step 1 The order in which two numbers are __________ does not change the sum. The property shown is the __________ Property of __________.

Step 2 Which number is not shown on the right side of the equation? Fill in the blank.

$12 + 9 = 9 + \_\_\_$

Step 3 Check. Add the numbers on each side of the equation.

$12 + 9 = 9 + \_\_\_$

$21 = \_\_\_$

Use the Commutative Properties to fill in each blank. Check your answer.

4 $3 \cdot 9 = \_\_\_ \cdot 3$

$27 = \_\_\_$

5 $\_\_\_ + 5 = 5 + 18$

$\_\_\_ = 23$

6 $8 + 11 = \_\_\_ + \_\_\_$

$\_\_\_ = \_\_\_$

7 $5 \cdot 6 = \_\_\_ \cdot \_\_\_$

$\_\_\_ = \_\_\_$

8 $4 + \_\_\_ = 7 + \_\_\_$

$\_\_\_ = \_\_\_$

9 $2 \cdot \_\_\_ = 6 \cdot \_\_\_$

$\_\_\_ = \_\_\_$

Draw a model to show each equation.

10 $2 \cdot 5 = 5 \cdot 2$

11 $7 + 3 = 3 + 7$

Step by Step Problem-Solving Practice

Problem-Solving Strategies

- ☐ Draw a diagram.
- ☐ Look for a pattern.
- ☐ Guess and check.
- ☑ Act it out.
- ☐ Solve a simpler problem.

Solve.

12 **SHOPPING** Jacob bought 3 boxes of pens with 5 pens in each box. Lydia bought 5 boxes of pens with 3 pens in each box. Compare the number of pens Jacob and Lydia bought. Justify your answer.

Understand Read the problem. Write what you know.

Jacob bought ______ boxes with ______ pens each.

Lydia bought ______ boxes with ______ pens each.

Plan Pick a strategy. One strategy is to act it out.

Arrange pens in rows and columns to show the pens that Jacob and Lydia bought. Then, write an expression to model each arrangement.

Solve Jacob's arrangement is ______ rows by ______ columns.

He has ______ pens.

The expression is ______.

Lydia's arrangement is ______ rows by ______ columns.

She has ______ pens.

The expression is ______.

The number of pens that Jacob bought is __________ the number of pens that Lydia bought.

Check Multiply the numbers on each side of the equation.

$3 \cdot 5 =$ ______ $\cdot$ ______

______ $=$ ______

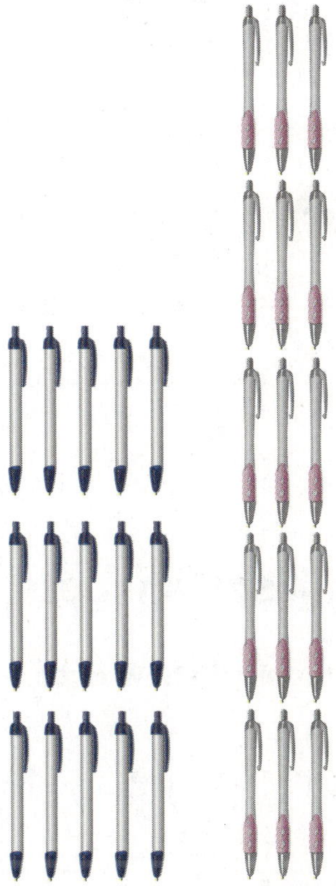

Jacob's pens　　Lydia's pens

GO ON

13 **JEWELRY** Missy has 7 boxes of necklaces with 3 necklaces in each box. Sari has 3 boxes of necklaces with 7 necklaces in each box. Compare the number of necklaces. Justify your answer. Check off each step.

____

_____ **Understand**

_____ **Plan**

_____ **Solve**

_____ **Check**

14 **COIN COLLECTING** Billy has 35 coins. His brother gave him 62 more coins. Omar has 62 coins. His father gave him 35 more coins. Compare the number of coins. Justify your answer.

____

15 **STORES** The Corner Store had 15 sweaters on display. Ten more sweaters were delivered in the afternoon. The Sweater Store had 10 sweaters on display. In the afternoon delivery, 15 more sweaters arrived. Compare the number of sweaters. Justify your answer.

____

16 **Reflect** How do you know that $65 + 98 = 98 + 65$ without adding?

____

____

____

Skills, Concepts, and Problem Solving

Draw a model to show each equation.

17 $6 \cdot 5 = 5 \cdot 6$
Which property did you show?

____

18 $4 + 5 = 5 + 4$
Which property did you show?

____

Use the Commutative Properties to fill in each blank. Check your answer.

19. $7 + 6 =$ ______ $+ 7$

______ $= 13$

20. $39 + 28 = 28 +$ ______

$67 =$ ______

21. $5 \times 9 = 9 \times$ ______

$45 =$ ______

22. $18 \times 5 =$ ______ $\times 18$

______ $= 90$

23. $48 + 37 =$ ______ $+$ ______

______ $=$ ______

24. ______ $+$ ______ $= 15 + 11$

______ $=$ ______

25. ______ $\cdot$ ______ $= 7 \cdot 6$

______ $=$ ______

26. $9 \cdot 8 =$ ______ $\cdot$ ______

______ $=$ ______

27. $5 +$ ______ $= 26 +$ ______

$31 =$ ______

28. ______ $+ 44 =$ ______ $+ 55$

______ $= 99$

29. $8 \cdot$ ______ $=$ ______ $\cdot 8$

$32 =$ ______

30. $5 \cdot$ ______ $= 9 \cdot$ ______

$45 =$ ______

31. $2 \cdot$ ______ $=$ ______ $\cdot 2$

______ $= 54$

32. ______ $\cdot 18 =$ ______ $\cdot 3$

______ $= 54$

Solve.

33. **HOBBIES** Sarah has 5 bags of marbles with 10 marbles in each bag. Noah has 10 bags of marbles with 5 marbles in each bag. Compare the number of marbles. Justify your answer.

____

34. **GAMES** Felipe had 14 game cards. He bought 12 more game cards. Wanda had 12 game cards. She bought 14. Compare the number of game cards. Justify your answer.

____

35. **MUSIC** Nicolas saved 23 songs to one jump drive and 35 to another. Eliza saved 35 songs to one jump drive and 23 to another. Compare the number of songs they saved. Justify your answer.

____

36. **NUMBERS** If you know that $195 + 126 = 321$, what is the sum of $126 + 195$? ______

GO ON

37 **NUMBERS** If you know that $26 \times 11 = 286$, what is the product of

11×26? ______

Vocabulary Check **Write the vocabulary word that completes each sentence.**

38 The Commutative Property of ____________ states that the order in which two numbers are added does not change the sum.

39 The Commutative Property of ____________ states that the order in which two numbers are multiplied does not change the product.

40 The answer to an addition problem is the ____________.

41 **Writing in Math** Use the Commutative Property of Multiplication to rewrite the expression $8 \cdot 7$. Does this change the product? Explain.

____

____

____

Spiral Review

Solve. (Lesson 1-4, p. 25)

42 **PHOTOGRAPHY** Yolanda took 180 digital pictures. She saved the pictures in folders on her laptop. She placed 36 pictures in each folder. How many folders did Yolanda create for her pictures?

______

43 **VOLUNTEERING** Maria and Rick volunteered at a local hospital. They set up 5 booths in 1 hour (60 minutes). How long did it take them to set up each booth?

____

44 Show that multiplying by 4 on each side of $(7 \cdot 8) = (28 \cdot 2)$ results in a true equation.

____

____

45 Show that adding 6 to each side of $(25 + 7) = (19 + 13)$ results in a true equation.

____

____

STOP

Lesson 2-2

Associative Property

2AF1.1 Use the commutative and associative rules to simplify mental calculations and to check results.
3AF1.5 Recognize and use the commutative and associative properties of multiplication.

KEY Concept

The Associative Property states that the way in which three numbers are grouped when they are added or multiplied does not change their sum or product.

Property	Definition	Example
Associative Property of Addition	$(a + b) + c = a + (b + c)$	$(4 + 6) + 9 = 4 + (6 + 9)$ $10 + 9 = 4 + 15$ $19 = 19$
Associative Property of Multiplication	$(a \cdot b) \cdot c = a \cdot (b \cdot c)$	$(2 \cdot 3) \cdot 5 = 2 \cdot (3 \cdot 5)$ $6 \cdot 5 = 2 \cdot 15$ $30 = 30$

VOCABULARY

Associative Property of Addition
the grouping of the *addends* does not change the *sum*

Associative Property of Multiplication
the grouping of the *factors* does not change the *product*

addend
any numbers or quantities being added together
(Lesson 1-1, p. 4)

factor
a number that divides a whole number evenly

The **Associative and Commutative Properties** can help you find sums and products mentally.

Example 1

Draw a model to show $(2 + 3) + 4 = 2 + (3 + 4)$. Which property did you show?

1. Create a model for each side of the equation.

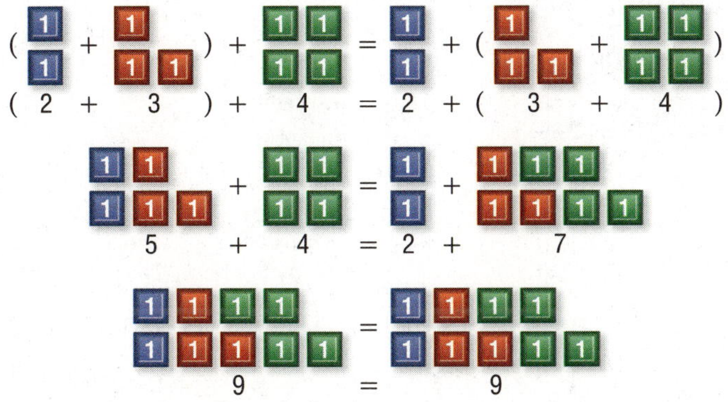

2. The grouping of the addends did not change the sum. This is the Associative Property of Addition.

YOUR TURN!

Draw a model to show $(1 \cdot 3) \cdot 2 = 1 \cdot (3 \cdot 2)$. Which property did you show?

1. Create a model for each side of the equation.

2. Which property did you show?

____

GO ON

Example 2

Use the Associative Property to fill in the blank. Check your answer.

$(7 \cdot 5) \cdot 2 = 7 \cdot (\_\_\_\_ \cdot 2)$

1. Use the Associative Property of Multiplication.
   $(7 \cdot 5) \cdot 2 = 7 \cdot (\underline{\ 5\ } \cdot 2)$

2. Check by multiplying the numbers on each side of the equation.
   $(7 \cdot 5) \cdot 2 = 7 \cdot (5 \cdot 2)$
   $35 \cdot 2 = 7 \cdot 10$
   $70 = 70$

YOUR TURN!

Use the Associative Property to fill in the blank. Check your answer.

$(2 + 4) + 1 = 2 + (\_\_\_\_ + 1)$

1. Use the Associative Property of Addition.
   $(2 + 4) + 1 = 2 + (\_\_\_\_ + 1)$

2. Check by adding the numbers on each side of the equation.

____

____

____

Who is Correct?

Use the Associative and Commutative Properties to find the sum of 25 + 94 + 75 mentally.

Add the numbers first that give you a 0 in the ones place. This will make it easier to mentally add the third number.

Danielle

$25 + 94 + 75$
$= (25 + 94) + 75$
$= 119 + 75$
$= 194$

Julie

$25 + 94 + 75$
$= 25 + 75 + 94$
$= (25 + 75) + 94$
$= 100 + 94$ or 194

Aaron

$25 + 94 + 75$
$= 25 + (94 + 75)$
$= 25 + 169$
$= 194$

Circle correct answer(s). Cross out incorrect answer(s).

Guided Practice

Draw a model to show each equation.

1. $(1 + 2) + 3 = 1 + (2 + 3)$
   Which property did you show?

____

2 $(2 \cdot 3) \cdot 4 = 2 \cdot (3 \cdot 4)$
Which property did you show?

____

____

Step by Step Practice

Use the Commutative and/or Associative Properties to find the sum mentally.

3 $8 + 4 + 16$

Step 1 Look for two numbers whose sum is 10, 20, 30, or another multiple of 10. The sum of ______ and ______ is a multiple of 10, ______.

Step 2 Rewrite the expression using the Associative Property.
$8 + 4 + 16 = 8 + (\_\_\_\_\_\_ + \_\_\_\_\_\_)$

Step 3 Find the sum.
$8 + (4 + 16) = 8 + \_\_\_\_\_\_$ or ______

Add the numbers first that make it easier to mentally add the third number.

Use the Associative Properties to fill in each blank. Check your answer.

4 $(9 \cdot 5) \cdot 2 = \_\_\_\_\_\_ \cdot (5 \cdot 2)$
$45 \cdot 2 = 9 \cdot \_\_\_\_\_\_$
$90 = \_\_\_\_\_\_$

5 $7 + (3 + 8) = (\_\_\_\_\_\_ + 3) + 8$
$7 + 11 = \_\_\_\_\_\_ + 8$
$18 = \_\_\_\_\_\_$

6 $5 \cdot 4 \cdot 3 = (\_\_\_\_\_\_ \cdot \_\_\_\_\_\_) \cdot \_\_\_\_\_\_$
$= \_\_\_\_\_\_ \cdot \_\_\_\_\_\_$
$= \_\_\_\_\_\_$

7 $(14 + 19) + 1 = \_\_\_\_\_\_ + (\_\_\_\_\_\_ + \_\_\_\_\_\_)$
$33 + 1 = \_\_\_\_\_\_ + \_\_\_\_\_\_$
$\_\_\_\_\_\_ = \_\_\_\_\_\_$

Use the Commutative and Associative Properties to find each sum or product mentally.

8 $9 + 21 + 30 = (\_\_\_\_\_\_ + \_\_\_\_\_\_) + \_\_\_\_\_\_$
$= \_\_\_\_\_\_ + \_\_\_\_\_\_$
$= \_\_\_\_\_\_$

9 $5 \cdot 21 \cdot 6 = \_\_\_\_\_\_ \cdot (\_\_\_\_\_\_ \cdot \_\_\_\_\_\_)$
$= \_\_\_\_\_\_ \cdot \_\_\_\_\_\_$
$= \_\_\_\_\_\_$

GO ON

Step by Step Problem-Solving Practice

Problem-Solving Strategies

- ☐ Look for a pattern.
- ☐ Guess and check.
- ☑ Solve a simpler problem.
- ☐ Work backward.

Solve.

10 **SCHOOL DAYS** Mr. Daniels sold 75 tickets to the school play on Monday, 52 tickets on Tuesday, and 48 tickets on Wednesday. How many tickets did Mr. Daniels sell in all? Explain your reasoning.

Understand Read the problem. Write what you know.
The number of tickets sold was ______, ______, and ______.

In this problem, the words "in all" mean to ______.

Plan Pick a strategy. One strategy is to solve a simpler problem.

Look for two numbers that will have a sum with a 0 in the ones place. Add those numbers together first.

Solve Write an expression for how many tickets Mr. Daniels sold in all.

______ + ______ + ______

Use the Associative Property of Addition to rewrite the expression so that it is easier to simplify mentally. Then find the sum.

______ + ______ + ______ = (______ + ______) + ______
= ______ + ______ = ______

Check Use subtraction to check your answer.

11 **STAMP COLLECTING** Uma has 44 stamps in her collection. Elena gave her 22 more stamps. Uma then bought 6 more stamps. How many stamps does Uma have in all? Explain your reasoning. Check off each step.

____

______ **Understand**

______ **Plan**

______ **Solve**

______ **Check**

12 **SPORTS** Marty bought the boxes of whistles shown. Each whistle cost \$2. How much did Marty spend? Explain your reasoning.

____

____

13 **COOKING** A casserole recipe calls for 2 packages of cheese. Selena needs to make 3 casseroles. Each package of cheese costs \$1.50. Find the cost of the chesse for all 3 casseroles. Explain your reasoning.

____

14 **Reflect** Give an example of the Associative Property of Addition. Check your answer.

____

____

Skills, Concepts, and Problem Solving

Draw a model to show each equation.

15 $(3 + 4) + 1 = 3 + (4 + 1)$
Which property did you show?

____

____

16 $(1 \cdot 5) \cdot 3 = 1 \cdot (5 \cdot 3)$
Which property did you show?

____

____

GO ON

Use the Associative Properties to fill in each blank. Check your answer.

17 $36 + (14 + 19) = (\_\_\_\_ + 14) + 19$

$36 + 33 = \_\_\_\_ + 19$

$\_\_\_\_ = \_\_\_\_$

18 $8 \cdot (5 \cdot 4) = (\_\_\_\_ \cdot 5) \cdot 4$

$8 \cdot 20 = \_\_\_\_ \cdot 4$

$\_\_\_\_ = \_\_\_\_$

19 $6 \cdot (5 \cdot 3) = \_\_\_\_\_\_\_\_\_\_\_\_$

$= \_\_\_\_ \cdot \_\_\_\_$

$= \_\_\_\_$

20 $18 + (22 + 37) = \_\_\_\_\_\_\_\_\_\_\_\_$

$= \_\_\_\_ + \_\_\_\_$

$= \_\_\_\_$

21 $(36 + 14) + 16 = \_\_\_\_\_\_\_\_\_\_\_\_$

$= \_\_\_\_ + \_\_\_\_$

$= \_\_\_\_$

22 $(63 + 13) + 17 = \_\_\_\_\_\_\_\_\_\_\_\_$

$= \_\_\_\_ + \_\_\_\_$

$= \_\_\_\_$

Use the Commutative and Associative Properties to find each sum or product mentally.

23 $15 \cdot 7 \cdot 2 = (\_\_\_\_ \cdot \_\_\_\_) \cdot \_\_\_\_$

$= \_\_\_\_ \cdot \_\_\_\_$

$= \_\_\_\_$

24 $12 \cdot 3 \cdot 5 = (\_\_\_\_ \cdot \_\_\_\_) \cdot \_\_\_\_$

$= \_\_\_\_ \cdot \_\_\_\_$

$= \_\_\_\_$

25 $102 + 89 + 18 = (\_\_\_\_ + \_\_\_\_) + \_\_\_\_$

$= \_\_\_\_ + \_\_\_\_$

$= \_\_\_\_$

26 $15 + 77 + 85 = (\_\_\_\_ + \_\_\_\_) + \_\_\_\_$

$= \_\_\_\_ + \_\_\_\_$

$= \_\_\_\_$

Solve. Justify your answer.

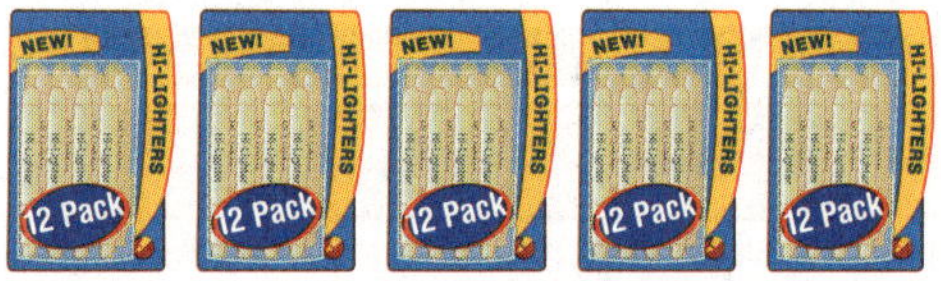

27 **SHOPPING** Sara bought the packages of highlighters shown. Each highlighter cost \$2. How much did Sara spend?

____

____

28 **CONSTRUCTION** Tregg has 17 nails. Isabel gave him 13 more nails. Tregg then bought 18 more nails. How many nails does Tregg have in all?

____

29 **FOOD** Diana needs to make 5 pumpkin pies for her school fundraiser. Each pumpkin pie needs 2 cans of pumpkin filling. Each can of pumpkin filling costs \$1.40. How much will the cans of pumpkin filling cost?

____

30 **NUMBERS** If you know that $3 + (5 + 4) = 12$, then what is the sum of $(3 + 5) + 4$? ______

31 **NUMBERS** If you know that $2 \times (6 \times 11) = 132$, what is the product of $(2 \times 6) \times 11$? ______

Vocabulary Check **Write the vocabulary word that completes each sentence.**

32 The Associative Property of ________________ states that the grouping of the factors does not change the product.

33 The Associative Property of ________________ states that the grouping of the addends does not change the sum.

34 **Writing in Math** Explain how you can use the Commutative and Associative Properties to help you find sums and products mentally.

____

____

____

Spiral Review

Solve. (Lesson 2-1, p. 42)

35 **SAFETY** Manuel has 5 boxes of safety glasses with 12 pairs of glasses in each box. Stephany has 12 boxes of safety glasses with 5 pairs of glasses in each box. Compare the number of glasses. Explain your reasoning.

____

Find the value of the variable by modeling each equation. (Lesson 1-4, p. 25)

36 $2 + t = 5$

$t =$ ______

37 $8 - \square = 3$

$\square =$ ______

STOP

Chapter 2

Progress Check 1 (Lessons 2-1 and 2-2)

Draw a model to show $1 + 5 = 5 + 1$. Which property did you show?

1 ______________________________

____

Draw a model to show $(1 \cdot 5) \cdot 3 = 1 \cdot (5 \cdot 3)$. Which property did you show?

2 ______________________________

____

Use the Commutative Property to fill in each blank with the correct value. Check your answer.

3 $21 + 36 =$ ______ + ______

______ = ______

4 $9 \cdot 3 =$ ______ • ______

______ = ______

Use the Commutative and Associative Properties to find each sum or product mentally.

5 $3 + 16 + 7 =$ (______ + ______) + ______

= ______ + ______ = ______

6 $5 \cdot 6 \cdot 3 =$ (______ • ______) • ______

= ______ • ______ = ______

Solve. Justify your answer.

7 **FITNESS** Brian does 30 minutes of aerobics on Tuesdays, 45 minutes on Thursdays, and 20 minutes on Saturdays. How many minutes of aerobics does Brian do altogether?

____

8 **SHOPPING** Lorenzo bought the packages of pens shown. Each pen cost $2. How much did Lorenzo spend?

____

Lesson 2-3

Distributive Property

5AF1.3 Know and use the distributive property in equations and expressions with variables.

KEY Concept

Distributive Property: The Distributive Property states that to multiply a sum by a number, you can multiply each addend by the number outside the parentheses. You can multiply a difference by a number in a similar way.

Definition	Example
$a(b + c) = ab + ac$	$2(4 + 5) = (2 \cdot 4) + (2 \cdot 5)$ $2 \cdot 9 = 8 + 10$ $18 = 18$
or	
$a(b - c) = ab - ac$	$3(7 - 4) = (3 \cdot 7) - (3 \cdot 4)$ $3(3) = 21 - 12$ $9 = 9$

VOCABULARY

Distributive Property of Multiplication
to multiply a *sum* by a number, you can multiply each *addend* by the number and add the *products*

Example 1

Use the Distributive Property and a model to find 5 • 12.

1. Draw a model to show 5 • 12.

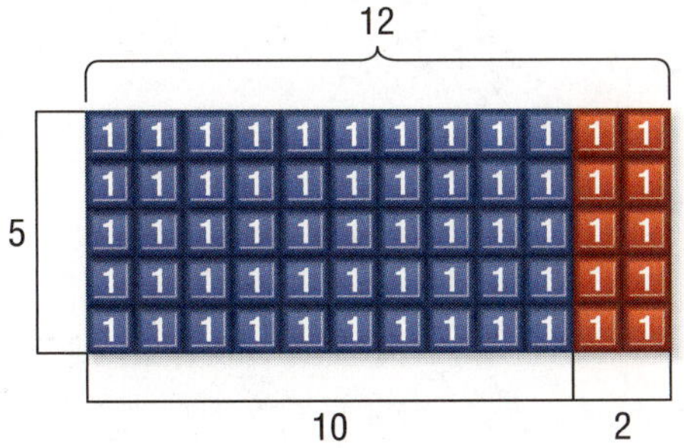

2. Draw a line to separate the factor 12 into tens and ones places.
   $5 \cdot 12 = 5 \cdot (10 + 2)$
3. Multiply to find the two products.
   $5 \cdot 10 = 50$
   $5 \cdot 2 = 10$
4. Add the products.
   $50 + 10 = 60$ So, $5 \cdot 12 = 60$.

YOUR TURN!

Use the Distributive Property and a model to find 3 • 13.

1. Draw a model to show 3 • 13.

2. Draw a line to separate the factor 13 into tens and ones places.

   ____

3. Multiply to find the two products.

   ____________ ____________

4. Add the products.

   ____

   ____

GO ON

Example 2

Use the Distributive Property to find $3(t - 2)$.

1. Use the Distributive Property. $3(t - 2) = (3 \cdot t) - (3 \cdot 2)$
2. Simplify inside the parentheses. $= 3t - 6$

YOUR TURN!

Use the Distributive Property to find $5(x + 1)$.

1. Use the Distributive Property. ____________________
2. Simplify inside the parentheses. ____________________

Who is Correct?

Use the Distributive Property to find $3(40 - 7)$.

Renee

$3(40 - 7) = 40 - (3 \times 7)$
$= 40 - 21$
$= 19$

Sunil

$3(40 - 7) = (3 \times 40) - 7$
$= 120 - 7$
$= 113$

Juan

$3(40 - 7) = (3 \times 40) - (3 \times 7)$
$= 120 - 21$
$= 99$

Circle correct answer(s). Cross out incorrect answer(s).

Guided Practice

1. Use the Distributive Property and a model to find $3 \cdot 14$.

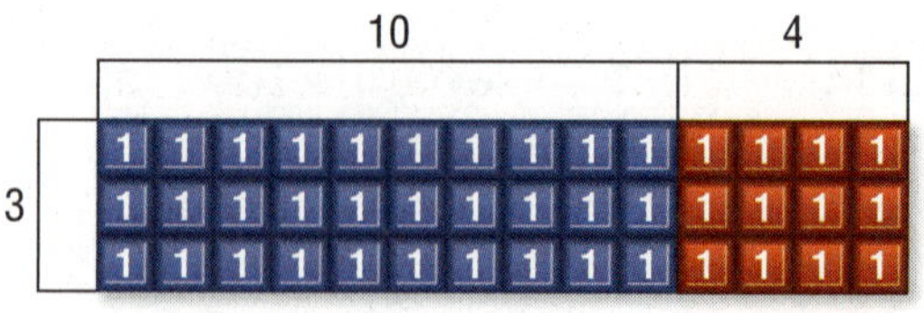

Multiply to find the two products.

____

Add the products.

____

____

Step by Step Practice

2 Use the Distributive Property to find $9(7 + 4)$.

Step 1 Use the Distributive Property.

$9(7 + 4) = (9 \cdot \_\_\_\_\_\_) + (9 \cdot \_\_\_\_\_\_)$

Step 2 Simplify inside the parentheses.

$(9 \cdot 7) + (9 \cdot 4) = \_\_\_\_\_\_ + \_\_\_\_\_\_$

Step 3 Add. ______ + ______ = ______

Use the Distributive Property to find each product. Show your work.

3 $6(6 - 4) = (\_\_\_\_\_\_ \cdot 6) - (\_\_\_\_\_\_ \cdot 4)$
$= \_\_\_\_\_\_ - \_\_\_\_\_\_$
$= \_\_\_\_\_\_$

4 $7(w + 5) = (\_\_\_\_\_\_ \cdot w) + (\_\_\_\_\_\_ \cdot 5)$
$= \_\_\_\_\_\_ + \_\_\_\_\_\_$
$= \_\_\_\_\_\_$

5 $8(5 + 6) =$ ______

6 $4(8 - a) =$ ______

Step by Step Problem-Solving Practice

Problem-Solving Strategies
- ☑ Use a model.
- ☐ Look for a pattern.
- ☐ Guess and check.
- ☐ Act it out.
- ☐ Solve a simpler problem.

Solve.

7 **BASKETBALL** Enrico made five 3-point baskets and Derek made six 3-point baskets in their last basketball game. How many points did they score in all?

Understand Read the problem. Write what you know.
Enrico made ______ 3-point baskets.
Derek made ______ 3-point baskets.

Plan Pick a strategy. One strategy is to use a model. Draw a model to represent the number of baskets the boys scored.

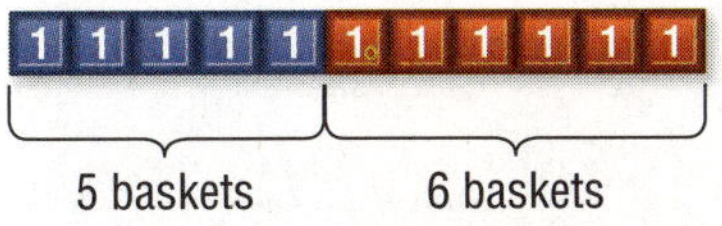

Solve Show the number of points scored.

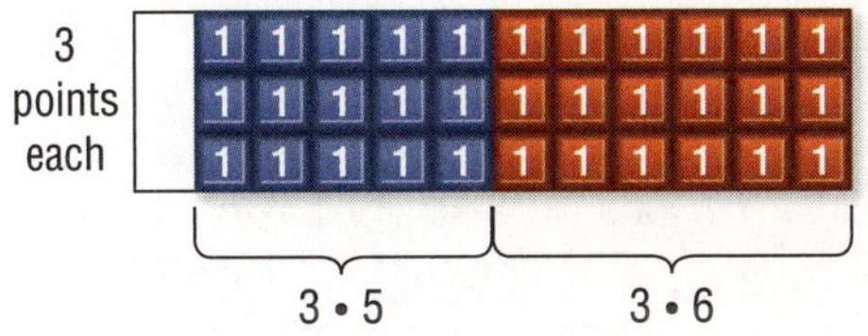

$\_\_\_\_\_\_ \cdot 5 + \_\_\_\_\_\_ \cdot 6$
$= \_\_\_\_\_\_ + \_\_\_\_\_\_$
$= \_\_\_\_\_\_$

Enrico and Derek scored ______ points in all.

Check Does your answer make sense? How many *baskets* were made in all? ______

GO ON

8 **SCHOOL** Miguel answered 12 five-point questions and 12 two-point questions correctly on his last history test. What was Miguel's test score? Check off each step.

____

_____ **Understand**

_____ **Plan**

_____ **Solve**

_____ **Check**

9 **MONEY** Rafiq earned \$15 each week for seven weeks. He spent \$6 each week for seven weeks. How much money does Rafiq have left?

____

10 **BASKETBALL** Penelope made 15 two-point baskets and 15 three-point baskets during this basketball season. How many points did Penelope score during the season?

____

11 **Reflect** What do you *distribute* when you use the Distributive Property?

____

____

Skills, Concepts, and Problem Solving

Use the Distributive Property and a model to find each product.

12 $4 \cdot 16 =$ _____

13 $5 \cdot 15 =$ _____

Write the value that makes each equation true.

14 $6 \times 27 = (6 \times 20) + (6 \times$ _____ $)$

15 $x(3 - 4) = (x \times$ _____ $) - (x \times 4)$

16 $15 \times 48 = (15 \times$ _____ $) + (15 \times 8)$

17 $28(8 - 5) = (28 \times$ _____ $) - (28 \times 5)$

Use the Distributive Property to find each product. Show your work.

18 $18(2 + 3) =$ ________

19 $3(8 + 13) =$ ________

20 $11(7 - 2) =$ ________

21 $4(12 - 5) =$ ________

22 $14(5 + 3a) =$ ________

23 $12(9 - 2g) =$ ________

24 $5(11 - x) =$ ________

25 $8(6 - y) =$ ________

26 $-2(w + 18) =$ ________

27 $-4(h + 15) =$ ________

Solve.

28 **FOOTBALL** The Mustangs scored 3 touchdowns (each worth 6 points) and 3 field goals (each worth 3 points). How many points did the football team score?

____

29 **MONEY** Marisa earned $18 each day for 8 days. She spent $3 on lunch each day for 8 days. How much money does Marisa have left?

____

30 **PARTIES** Your parents have agreed to pay for a party at your favorite restaurant. There will be 20 people at the party. Use the menu. How much will your parents spend if everyone orders the hamburger/fries special with a large soda?

____

31 **TRIPS** You are part of a group of 20 friends planning a trip to the art museum. Suppose admission to the museum costs $6 and a bus ticket to the museum costs $2.50. What is the total cost of the trip?

____

GO ON

Vocabulary Check **Write the vocabulary word that completes the sentence.**

32 The ______________ Property states that to multiply a sum by a number, you can multiply each addend by the same number and add the products.

33 **Writing in Math** The Distributive Property can be used to write $5 \times (4 + 4) = (5 \times 4) + (5 \times 4)$. Can the Distributive Property also be used to write $5 \times (7 + 1) = (5 \times 7) + (5 \times 1)$? Explain why or why not.

____

____

Spiral Review

Use the Commutative and Associative Properties to find each sum or product mentally. (Lesson 2-2, p. 49)

34 $12 + 37 + 18 = (\_\_\_\_ + \_\_\_\_) + \_\_\_\_$

$= \_\_\_\_ + \_\_\_\_$

$= \_\_\_\_$

35 $8 \cdot 7 \cdot 5 = \_\_\_\_ \cdot (\_\_\_\_ \cdot \_\_\_\_)$

$= \_\_\_\_ \cdot \_\_\_\_$

$= \_\_\_\_$

Solve. (Lesson 1-2, p. 11)

36 **BOOKS** Fala has 15 books. She separated the books equally among her 3 friends. How many books did each friend get?

____

Name the operation needed to solve the problem. Write a number sentence to solve the problem. (Lesson 1-1, p. 4)

37 **SHOPPING** Elan spent the day shopping on Saturday. Use his receipts to determine the total amount he spent.

____

STOP

Lesson 2-4

Order of Operations

7AF1.2 Use the correct order of operations to evaluate algebraic expressions such as $3(2x + 5)^2$.

KEY Concept

You must follow the **order of operations** to evaluate mathematical expressions correctly.

Order of Operations	Symbol
1. Simplify grouping symbols.	(parentheses) [brackets] $\frac{2-1}{5-3}$ fraction bar
2. Simplify exponents.	base → 2^5 ← exponent
3. Multiply and divide in order from left to right.	× • ÷ /
4. Add and subtract in order from left to right.	+ −

VOCABULARY

exponent
the number of times a *base* is multiplied by itself
Example: In 2^5, 5 is the exponent.

base (of a power)
the number used as the factor
Example: In 2^5, 2 is the base.

order of operations
rules that tell what order to use when evaluating expressions
(1) Simplify grouping symbols.
(2) Simplify exponents.
(3) Multiply and divide in order from left to right.
(4) Add and subtract in order from left to right.

Sometimes parentheses are used to set a number apart from other operations. If there is no operation to be performed inside the parentheses, check for **exponents**.

Example 1

Find the value of 3 − 2 + 12 ÷ 4.

Use the order of operations. There are no grouping symbols or exponents.

$3 - 2 + 12 \div 4 = 3 - 2 + 3$ Multiply and divide from left to right.
$= 1 + 3$ Add and subtract from left to right.
$= 4$

From left to right, subtraction comes first in this expression.

YOUR TURN!

Find the value of 7 − 5 + 3 × 3.

Use the order of operations. There are no grouping symbols or exponents.

$7 - 5 + 3 \times 3 = 7 - 5 +$ ______
= ______ + ______
= ______

GO ON

Example 2

Find the value of $58 - \frac{36 - 12}{6 + 2} \times 3^2$.

$$58 - \frac{36 - 12}{6 + 2} \times 3^2 = 58 - \frac{24}{8} \times 3^2 \quad \text{Simplify grouping symbols.}$$

$$= 58 - \frac{24}{8} \times 9 \quad \text{Simplify exponents.}$$

$$= 58 - 3 \times 9 \quad \text{Divide.}$$

$$= 58 - 27 \quad \text{Multiply.}$$

$$= 31 \quad \text{Subtract.}$$

YOUR TURN!

Find the value of $20 \div 5 + (3 + 2)^2 \cdot 2 - 6$.

$20 \div 5 + (3 + 2)^2 \cdot 2 - 6 = 20 \div 5 +$ ______ $\cdot 2 - 6$

$= 20 \div 5 +$ ______ $\cdot 2 - 6$

$= 4 +$ ______ $- 6$

$=$ ______ $- 6$

$=$ ______

Who is Correct?

Find the value of $10 \div 2 + (2 + 2)^2 \cdot 2$.

Dontel

$10 \div 2 + (2 + 2)^2 \cdot 2$
$= 10 \div 2 + 4^2 \cdot 2$
$= 5 + 16 \cdot 2$
$= 21 \cdot 2$
$= 42$

Corey

$10 \div 2 + (2 + 2)^2 \cdot 2$
$= 5 + (4 + 4) \cdot 2$
$= 5 + (8) \cdot 2$
$= 5 + 16$
$= 21$

Rachel

$10 \div 2 + (2 + 2)^2 \cdot 2$
$= 10 \div 2 + 4^2 \cdot 2$
$= 10 \div 2 + 16 \cdot 2$
$= 5 + 32$
$= 37$

Circle correct answer(s). Cross out incorrect answer(s).

Guided Practice

Name the step that should be performed first in each expression.

1. $4 \cdot 6 + (30 - 3) \div 3^2$ ______________________
2. $26 \div 1 - (1 + 7) \cdot 2$ ______________________
3. $9 + 5^2 \div 5 + 2 \cdot 6$ ______________________
4. $8 + 2 \div 2 \cdot 5 - 1$ ______________________

Step by Step Practice

5 Find the value of $14 - 6 \cdot (2 - 1)^2 - 5 + 2$.

Step 1 Use the order of operations. Simplify the grouping symbols.

$14 - 6 \cdot (2 - 1)^2 - 5 + 2 = 14 - 6 \cdot (\_\_\_\_)^2 - 5 + 2$

Step 2 Simplify the exponent.

$14 - 6 \cdot 1^2 - 5 + 2 = 14 - 6 \cdot \_\_\_\_ - 5 + 2$

Step 3 Multiply and divide.

$14 - 6 \cdot 1 - 5 + 2 = 14 - \_\_\_\_ - 5 + 2$

Step 4 Add and subtract.

$14 - 6 - 5 + 2 = \_\_\_\_ - 5 + 2$

$= \_\_\_\_ + 2$

$= \_\_\_\_$

Find the value of each expression.

6 $9 - (1 + 8) + 3 \cdot 3^2 = 9 - \_\_\_\_ + 3 \cdot 3^2$

$= 9 - \_\_\_\_ + 3 \cdot \_\_\_\_$

$= 9 - \_\_\_\_ + \_\_\_\_$

$= \_\_\_\_ + \_\_\_\_$

$= \_\_\_\_$

7 $8 \div 2 + (5 \cdot 2)^2 - 7 = 8 \div 2 + (\_\_\_\_)^2 - 7$

$= 8 \div 2 + \_\_\_\_ - 7$

$= \_\_\_\_ + \_\_\_\_ - 7$

$= \_\_\_\_ - 7$

$= \_\_\_\_$

8 $20 - 4^2 \div 4 \cdot 2 + (7 - 4) = \_\_\_\_$

9 $(4 - 3)^2 \cdot 50 \div 5 - (3 + 6) = \_\_\_\_$

10 $30 \div \frac{21 + 4}{6 - 1} \div 2 \cdot 12 = \_\_\_\_$

11 $50 \div (7 + 3) \cdot 4 \div 2 = \_\_\_\_$

12 $(12^2 - 2^2) \div (6 + 1) - 8 = \_\_\_\_$

13 $[15^2 - 3^2 - (2 \cdot 8)] \div 25 = \_\_\_\_$

GO ON

Step by Step Problem-Solving Practice

Problem-Solving Strategies
- ☐ Draw a diagram.
- ☐ Guess and check.
- ☐ Act it out.
- ☑ Solve a simpler problem.
- ☐ Work backward.

14 **NUTRITION** Marcos has 2 baskets that hold 12 oranges each. He has 6 more baskets of 10 oranges each. Write and simplify an expression to find how many oranges Marcos has in all.

Understand Read the problem. Write what you know.
There are baskets with ______ oranges each
and baskets with ______ oranges each.

Plan Pick a strategy. One strategy is to solve a simpler problem. In this case, solving a simpler problem means to work on smaller parts of the expression, one at a time.

Solve Write an expression for the total number of oranges.

$\_\_\_\_ \cdot 12 \quad + \quad \_\_\_\_ \cdot 10$

2 baskets of 12   6 baskets of 10

Simplify the expression using the order of operations.

$2 \cdot 12 + 6 \cdot 10$

$= \_\_\_\_ + \_\_\_\_$ Multiply.

$= \_\_\_\_$ Add.

Marcos has ______ oranges.

Check You can use addition to check.
$12 + 12 + 10 + 10 + 10 + 10 + 10 + 10 = \_\_\_\_$

Write and simplify an expression to solve each problem.

15 **NATURE** Brandy likes to watch birds. She saw a wren make 2 nests. Three wrens each made 4 nests. Five nests were damaged during a storm. How many nests were left? Check off each step.

____

______ **Understand**

______ **Plan**

______ **Solve**

______ **Check**

16 **COLLECTIONS** Tyler bought 3 packs of comic books. Each pack had 5 comic books. He gave 7 comic books to his brother. Then he bought 2 more packs of comic books with 18 books in each. How many comic books does Tyler have now?

____

17 **PHOTOGRAPHY** Marlene was looking through her photo album. She looked at 6 pages with 4 photos each. She removed 2 photos. Then she looked at 10 pages with 8 photos each. She removed 6 photos. Finally she looked at 2 pages with 2 photos each. How many photos were left in the album?

____

____

18 **Reflect** Explain why $16 \div 2 + 6$ has a different value than $16 \div (2 + 6)$.

____

____

Skills, Concepts, and Problem Solving

Name the step that should be performed first in each expression.

19 $3 \cdot 2 + (4 \div 2) - 2^2$ ______

20 $2 \cdot (3 - 6)^2 + 9 \div 3$ ______

21 $(7 - 2^2 \cdot 4) - 8 \div 1$ ______

22 $4 + (6 - 2 \cdot 7) \div 2^2$ ______

23 $3[(100 + 25) \times 2] - 25$ ______

24 $\dfrac{18 + 66}{35 - 14}$ ______

Find the value of each expression.

25 $50 \div 5 + 3 \cdot 2^2 - (8 - 2) =$ ______

26 $3^2 + 8 \div 2 - (10 + 2) =$ ______

27 $16 - 4^2 \cdot 0 + 18 - 15 =$ ______

28 $(9 - 6)^2 + 8 \div 4 + 5 \cdot 6 =$ ______

29 $\dfrac{15 + 35}{21 + 4} \times 5 =$ ______

30 $\dfrac{2(14 - 6)}{2^2} =$ ______

31 $10[8(15 - 7) - (4 \times 3)] =$ ______

32 $5[(12 + 5) - 3(19 - 14)] =$ ______

GO ON

Write and simplify an expression to solve each problem.

33 **BOOKS** Ramona borrowed 2 stacks of books with 8 books each. She then returned 9 books. Then Ramona borrowed 2 stacks of 5 books each. How many books does Ramona have now?

____

34 **COLLECTIONS** Don had 100 collector cards. He sold 5 packs of baseball cards with 10 cards each. He then bought 3 packs of football cards with 12 cards each. Then Don sold 25 hockey cards. How many cards does Don have left?

____

____

Vocabulary Check **Write the vocabulary word that completes each sentence.**

35 A(n) ______________ is the number of times a base is multiplied by itself.

36 The ______________ is a set of rules that tells what order to follow when evaluating an expression.

37 **Writing in Math** Does 30 – (10 – 5) equal (30 – 10) – 5? Explain.

____

____

Spiral Review

Solve. Explain your reasoning. (Lesson 2-2, p. 49)

38 **SPORTS** Amy bought the boxes of softballs shown. Each softball cost $4. How much did Amy spend?

____

Use the Commutative and Associative Properties fo fill in each blank. Check your answer. (Lesson 2-1, p. 42)

39 68 + 16 = ______ + ______

______ = ______

40 ______ + ______ = 12 + 17

______ = ______

STOP

Chapter 2

Progress Check 2 (Lessons 2-3 and 2-4)

Use the Distributive Property and a model to find each product.

1. $4 \cdot 13 =$ ______

2. $6 \cdot 16 =$ ______

Name each operation that should be performed first.

3. $8 - 4 \cdot (7 + 4)^2 \div 2$ ______

4. $3^2 \cdot 2 - (12 \div 4) + 6$ ______

Use the Distributive Property to find each product. Show your work.

5. $7(3 + 2) =$ ______

6. $6(8 - 2) =$ ______

7. $3(k + 5) =$ ______
______

8. $-6(p - 7) =$ ______
______

Find the value of each expression.

9. $18 + 2^2 \div 4 \cdot (5 - 2) + 7 =$ ______

10. $10 - (2 - 1)^2 + 16 \div 2 \cdot (1 + 1) =$ ______

11. $28 \div 2^2 \cdot 8 + 4 \div 2 =$ ______

12. $64 \div 4^2 \cdot 25 - (30 - 18) \div 4 =$ ______

Solve. Explain your reasoning.

13. **BASKETBALL** Tamar made six 2-point field goals and two 3-point field goals. How many points did Tamar score?

______

14. **SHOPPING** Payton had 50 pencils. He sold 3 bags of pencils with 5 pencils each. He then bought 2 packs of pencils with 10 pencils each. Then Payton gave 20 pencils to his sister. How many pencils does Payton have left?

______

______

Chapter 2

Study Guide

Vocabulary and Concept Check

Associative Property of Addition, *p. 49*

Associative Property of Multiplication, *p. 49*

base (of a power), *p. 63*

Commutative Property of Addition, *p. 42*

Commutative Property of Multiplication, *p. 42*

Distributive Property, *p. 57*

exponent, *p. 63*

order of operations, *p. 63*

Write the vocabulary word that completes each sentence.

1. The property that states that the order in which two numbers are multiplied does not change the *product* is the ______________________________.

2. The property that states that the grouping of the factors does not change the *product* is the ______________________________.

3. $(2 + 4) + 7 = 2 + (4 + 7)$ shows the ______________________________.

4. $a(b + c) = ab + ac$ or $a(b - c) = ab - ac$ is the ______________________________.

5. $3 + 9 = 9 + 3$ shows the ______________________________.

6. The ______________________________ is a set of rules that tells what order to use when evaluating an expression.

Write the correct vocabulary term in each blank.

7. ____________________ (arrow to the 4 in) 4^3

8. ____________________ (arrow to the 5 in) 2^5

9. ____________________

1) Simplify within parentheses.

2) Simplify exponents.

3) Multiply and divide from left to right.

4) Add and subtract from left to right.

Lesson Review

2-1 Commutative Property (pp. 42–48)

Use the Commutative Properties to fill in each blank. Check your answer.

10 $2 \cdot 8 =$ ______ $\cdot\ 2$

$16 =$ ______

11 ______ $+ 7 = 7 + 15$

______ $=$ ______

12 ______ $+ 4 =$ ______ $+$ ______

______ $= 22$

Example 1

Draw a model to show 4 + 2 = 2 + 4. Which property did you show?

1. Create a model for each side of the equation.

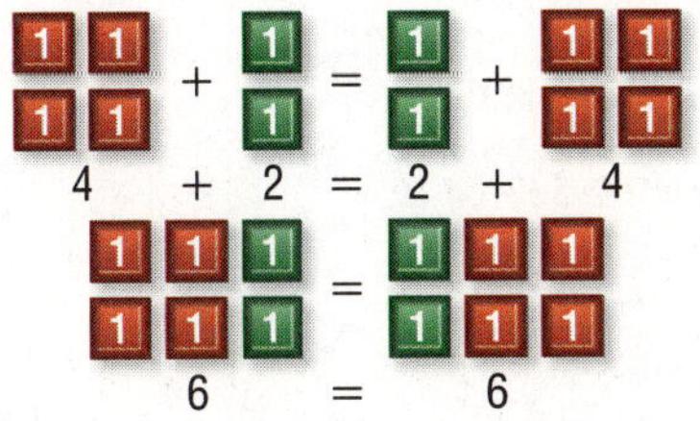

2. The order of the numbers changed, but the sum did not. This is the Commutative Property of Addition.

2-2 Associative Property (pp. 49–55)

Use the Associative Properties to fill in each blank. Check your answer.

13 $8 \cdot (4 \cdot 3) = (8 \cdot$ ______ $) \cdot 3$

____

____

14 $(5 + 9) + 3 = 5 + ($ ______ $+ 3)$

____

____

15 $2 + (13 + 15) = ($ ______ $+$ ______ $) +$ ______

$2 +$ ______ $= 15 +$ ______

______ $=$ ______

Example 2

Use the Associative Property to fill in each blank. Check your answer.

$(6 \cdot 3) \cdot 2 = 6 \cdot ($ ______ $\cdot\ 2)$

Use the Associative Property of Multiplication.

$(6 \cdot 3) \cdot 2 = 6 \cdot (\underline{\ 3\ } \cdot 2)$

Check by multiplying the numbers on each side of the equation.

$(6 \cdot 3) \cdot 2 = 6 \cdot (3 \cdot 2)$

$18 \cdot 2 = 6 \cdot 6$

$36 = 36$

Draw a model to show each equation.

16 $(2 \cdot 3) \cdot 5 = 2 \cdot (3 \cdot 5)$
Which property did you show?

____

____

17 $(2 + 3) + 4 = 2 + (3 + 4)$
Which property did you show?

____

Example 3

Draw a model to show
$(1 \cdot 3) \cdot 4 = 1 \cdot (3 \cdot 4)$.
Which property did you show?

1. Create a model for each side of the equation.

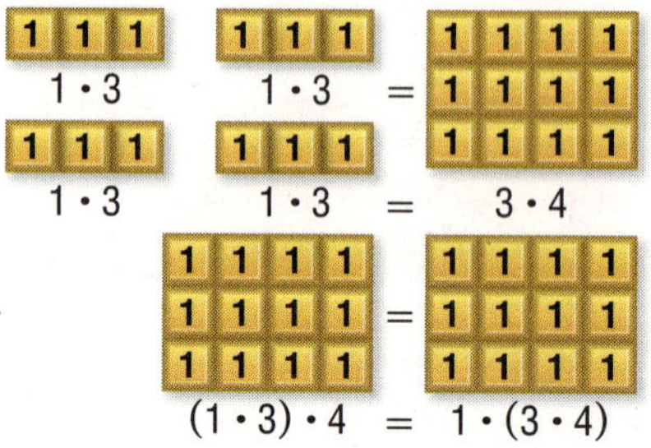

2. The grouping of the factors did not change the product.

Associative Property of Multiplication

Use the Commutative and Associative Properties to find each sum or product mentally.

18 $4 \cdot 7 \cdot 5 = (4 \cdot \_\_\_\_) \cdot 7$
$= \_\_\_\_ \cdot 7$
$= \_\_\_\_$

19 $11 + (9 + 4) = (\_\_\_\_ + 9) + 4$
$= \_\_\_\_ + 4$
$= \_\_\_\_$

Example 4

Use the Commutative and Associative Properties to find the sum mentally.

$15 + 9 + 5$

Determine which grouping would help you find the sum using mental math.
Group 15 and 5. Then find the sum.

$15 + 9 + 5 = 15 + 5 + 9$
$= (15 + 5) + 9$
$= 20 + 9$ or 29

2-3 Distributive Property (pp. 57–62)

Use the Distributive Property to find each product.

20 $9(5 - 3) = (\_\_\_\_ \cdot 5) - (\_\_\_\_ \cdot 3)$
$= \_\_\_\_ - \_\_\_\_$
$= \_\_\_\_$

21 $5(x + 8) = (\_\_\_\_ \cdot x) + (\_\_\_\_ \cdot 8)$
$= \_\_\_\_ + \_\_\_\_$
$= \_\_\_\_$

Example 5

Use the Distributive Property to find 7(8 + 3).

$7(8 + 3) = (7 \cdot 8) + (7 \cdot 3)$

Simplify within the parentheses. Then add.
$(7 \cdot 8) + (7 \cdot 3) = 56 + 21$
$= 77$

2-4 Order of Operations (pp. 63–68)

Find the value of each expression.

22 $10 + 8 \div 2 - 1 \times 3 = \_\_\_\_$

23 $11 - 7 + 3 \times 8 = \_\_\_\_$

24 $125 \div 5^2 \cdot 7 = \_\_\_\_$

Example 6

Find the value of 6 − 2 + 15 ÷ 3 × 4.

Use the order of operations.
There are no grouping symbols or exponents.
Multiply and divide.
$6 - 2 + 15 \div 3 \times 4 = 6 - 2 + 5 \times 4$
$= 6 - 2 + 20$

Add and subtract.
$6 - 2 + 20 = 4 + 20$
$= 24$

Find the value of each expression.

25 $3 \times 4 - (4 - 2)^3 + 6 \div 2 = \_\_\_\_$

26 $3^2 - 4 + (1 + 2)^3 \div 9 = \_\_\_\_$

27 $6^2 + 4 \div (8 - 6) \times 2 = \_\_\_\_$

Example 7

Find the value of $18 \div 3 + (2 + 1)^2 \cdot 4 - 5$.

Use the order of operations.
Simplify the grouping symbols.
$18 \div 3 + (2 + 1)^2 \cdot 4 - 5 = 18 \div 3 + 3^2 \cdot 4 - 5$

Simplify the exponent.
$18 \div 3 + 3^2 \cdot 4 - 5 = 18 \div 3 + 9 \cdot 4 - 5$

Multiply and divide.
$18 \div 3 + 9 \cdot 4 - 5 = 6 + 36 - 5$

Add and subtract.
$6 + 36 - 5 = 42 - 5$
$= 37$

Chapter 2

Chapter Test

Use the Commutative Properties to fill in each blank. Check your answer.

1 $4 \cdot 6 =$ ______ $\cdot\ 4$

$24 =$ ______

2 ______ $+ 9 = 9 + 11$

______ $=$ ______

3 $7 + 9 =$ ______ $+$ ______

______ $=$ ______

4 $7 \cdot 12 =$ ______ $\cdot$ ______

______ $=$ ______

5 Give an example of the Commutative Property of Addition. Check your example.

____

Use the Associative Properties to fill in each blank. Check your answer.

6 $3 \cdot (4 \cdot 10) = ($ ______ $\cdot\ 4) \cdot 10$

$=$ ______ $\cdot\ 10$

$=$ ______

7 $9 + (1 + 15) = ($ ______ $+ 1) + 15$

$=$ ______ $+ 15$

$=$ ______

8 $4 + (16 + 13) = ($ ______ $+$ ______ $) +$ ______

$=$ ______ $+$ ______

$=$ ______

9 $(7 \cdot 2) \cdot 5 =$ ______ $\cdot\ ($ ______ $\cdot$ ______ $)$

$=$ ______ $\cdot$ ______

$=$ ______

10 Give an example of the Associative Property of Multiplication. Check your answer.

____

Use the Distributive Property to find each product. Show your work.

11 $9(2 + 5) =$ ______________

12 $2(20 - y) =$ ______________

Find the value of each expression.

13 $16 + 2^3 \div 4 \cdot 5 - (16 - 10) =$ ______

14 $(7 - 3)^2 \cdot 3 \div 4 + (11 + 8) =$ ______

GO ON

Solve. Explain your reasoning.

15 **COOKING** Hoshi needs 3 gallons of stew for a potluck dinner. Each gallon of stew requires 2 cans of beef gravy. Each can of beef gravy costs $1.20. How much will the gravy for the stew cost?

____

____

16 **SHOPPING** Devin bought 4 boxes of markers with 5 markers in each box. Jennifer purchased 5 boxes with 4 markers in each box. Compare the number of markers.

____

17 **FITNESS** Sofia participated in two 50-minute aerobics sessions last week and three 50-minute aerobics sessions this week. How many minutes did she work out during both weeks?

____

18 **POPULATION** An apartment complex has 3 units. Four people lived in each unit. Then 8 people moved away. The next month 2 families of 5 moved into the complex. What is the total number of people living in the apartment complex now?

____

Correct the mistakes.

19 **ART** Terri purchased 3 boxes of acrylic paint tubes, each containing 8 tubes of paint. Raul purchased 8 boxes, each containing 3 acrylic paint tubes. Raul told Terri that he bought more tubes of paint than she did because he bought more boxes. What mistake did Raul make?

____

____

20 **FOOD** At the school cafeteria, the cook had 7 sandwiches. Two students purchased 1 each. The cook then sold two teachers 2 each. Her assistant made 11 more sandwiches. She dropped 1 of them on the floor so it had to be thrown away. The assistant said, "Now you only have 10 sandwiches to sell." What mistake did she make?

____

____

STOP

Chapter 2 Standards Practice

Choose the best answer and fill in the corresponding circle on the sheet at right.

1 If $9 \times 8 \times 7 = 504$, then what is $7 \times 9 \times 8$?

A 567

B 504

C 448

D 343

2 $49 \times (130 \times 62) =$

F $(48 \times 130) \times 62$

G $(49 \times 130) \times 62$

H $(49 \times 103) \times 62$

J $(47 \times 130) \times 62$

3 Olivia wants to paint the two opposite walls in her bedroom. If she knows the dimensions of one wall, which expression can help her figure out the square footage of both walls?

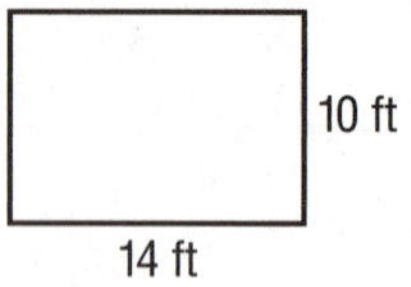

A $2(10 \times 14)$

B $2(14) + 10$

C $2(14) \times 2(10)$

D $2 + (14 \times 10)$

4 $234 \div 3 \times [5^2 - (4 \times 3)] =$

F 1,014

G 954

H 27

J 6

5 Which property is shown in the sentence below?

$(21 \times 8) + (21 \times 5) = 21 \times (8 + 5)$

A Associative Property of Addition

B Distributive Property of Multiplication over Addition

C Commutative Property of Addition

D Identity Property of Multiplication

6 Which property is shown in the sentence below?

$(346 \times 751) \times 203 = 346 \times (751 \times 203)$

F Associative Property of Multiplication

G Distributive Property of Multiplication over Addition

H Commutative Property of Multiplication

J Identity Property of Multiplication

7 Which property is shown in the sentence below?

$33.7 \times 1.8 = 1.8 \times 33.7$

A Associative Property of Addition

B Distributive Property of Multiplication over Addition

C Commutative Property of Multiplication

D Identity Property of Multiplication

GO ON

8 Gabriel buys 6 books at a back-to-school sale. Which expression has a value equal to the cost of 6 books?

Item	Cost
CD	$15
Book	$24
T-shirt	$19

F $(2 \times 24) + (3 \times 24)$

G $(2 \times 15) + (4 \times 15)$

H $(2 \times 19) + (3 \times 19)$

J $(2 \times 24) + (4 \times 24)$

---

9 Which symbol makes the sentence true?

$$216 \square 4 = 54$$

A $+$

B $-$

C $\times$

D $\div$

---

10 Which shows the difference of a number and 46 is 108?

F $x + 46 = 108$

G $x - 46 = 108$

H $x \times 46 = 108$

J $x \div 46 = 108$

---

11 Which is the answer to a division statement?

A difference

B product

C quotient

D sum

12 How many pounds of bananas can Andre buy with $3.00?

F 7 pounds

G 8 pounds

H 9 pounds

J 10 pounds

ANSWER SHEET

Directions: Fill in the circle of each correct answer.

1	Ⓐ	Ⓑ	Ⓒ	Ⓓ
2	Ⓕ	Ⓖ	Ⓗ	Ⓙ
3	Ⓐ	Ⓑ	Ⓒ	Ⓓ
4	Ⓕ	Ⓖ	Ⓗ	Ⓙ
5	Ⓐ	Ⓑ	Ⓒ	Ⓓ
6	Ⓕ	Ⓖ	Ⓗ	Ⓙ
7	Ⓐ	Ⓑ	Ⓒ	Ⓓ
8	Ⓕ	Ⓖ	Ⓗ	Ⓙ
9	Ⓐ	Ⓑ	Ⓒ	Ⓓ
10	Ⓕ	Ⓖ	Ⓗ	Ⓙ
11	Ⓐ	Ⓑ	Ⓒ	Ⓓ
12	Ⓕ	Ⓖ	Ⓗ	Ⓙ

Success Strategy

If two answers seem correct, compare them for differences. Reread the problem to find the best answer between the two.

STOP

Index